# BIOLOGY THROUGH THE EYES OF FAITH

OTHER BOOKS IN THIS SERIES:

*Business Through the Eyes of Faith*

*History Through the Eyes of Faith*

*Literature Through the Eyes of Faith*

*Psychology Through the Eyes of Faith*

*Sociology Through the Eyes of Faith*

# BIOLOGY THROUGH THE EYES OF FAITH

Richard T. Wright

HarperSanFrancisco
*A Division of* HarperCollins*Publishers*

The Christian College Coalition is an association of Christian liberal arts colleges and universities across North America. More than 30 Christian denominations, committed to a variety of theological traditions and perspectives, are represented by our member colleges. The views expressed in this volume are primarily those of the author(s) and are not intended to serve as a position statement of the Coalition membership.

For more information please contact the Christian College Coalition at 329 Eighth Street NE, Washington, DC 20002.



**Library of Congress Cataloging-in-Publication Data**

Wright, Richard T.
Biology through the eyes of faith.
1. Creation. 2. Biology—Religious aspects—Christianity. 3. Religion and science—1946– 4. Evolution. I. Title.
BS651.W84 1989 215'.74 88-45686
ISBN 0-06-069695-8 (pbk.)

94 95 96 RRD-H 12 11 10 9 8 7 6

## BIOLOGY TASK FORCE MEMBERS

Susan P. Bratton
U.S. National Park Service
University of Georgia

Calvin B. DeWitt
Au Sable Environmental Institute
University of Wisconsin

Robert L. Herrmann
American Scientific Affiliate

Martin LeBar
Central Wesleyan College

Max R. Terman
Tabor College

L. Duane Thurman
Oral Roberts University

David L. Wilcox
Eastern College

## SERIES ADVISORY BOARD MEMBERS

# CONTENTS

# FOREWORD

Someone committed to the Christian faith who reads contemporary biology will find many questions coming to mind. This book is an address to some of the most fundamental of those questions, written by an eminently competent research biologist working in the mainstream of contemporary biology. Dr. Wright poses the questions: What does biology look like when seen through the eyes of faith? What illumination and guidance does faith give us in our pursuit of, and reflections on, biology? What illumination does biology give us on faith? What are the points of tension on which continuing reflection is called for?

There are of course other books on the market that talk about the relation of faith to biology. Some of those propound the thesis that these must be seen as separate phenomena concerned with distinct domains of reality. Others affirm that we must see between these not isolation but opposition. Some of these latter argue that biology must stay and Christianity go. Perhaps more popular now are the books of the creation scientists who argue that Christianity must stay and contemporary biology go. They propose replacing it with something that they call "creation science."

Professor Wright does not embrace the isolation thesis. He is convinced that the Christian scriptures speak to that very same reality which the biologist studies. But neither does he suggest that either Christianity or contemporary biology must be discarded. He regards contemporary biology as giving a well-grounded, though indeed fallible and tentative, explanation of the biological world. But also he lives and works out of the conviction that Christianity is true. What he offers us here, then, is not mighty slashes with a blunt sword but a careful, thoughtful, detailed approach to the issues.

Two things especially are impressive about the book. One is its comprehensiveness—and that in two dimensions. Professor Wright develops an exceptionally comprehensive approach to biology, while at the same time his theology is both broad and deep. The result of bringing these together is illuminating and stimulating.

What also impresses one is the honesty and seriousness of the book. There are no hard questions that Professor Wright dodges. He faces up to them. And he always treats the issues with care and depth. Where he does not know, he says that. Where the evidence is not decisive, he tells us. Where others make do with glib slogans or quick solutions, Professor Wright offers careful and honest analysis.

This book, then, is the outcome of the honest struggle of a devout Christian and skilled biologist to arrive at wholeness, integrity—to arrive at the point where he sees how his faith and his biology fit together. You will find it a struggle worth participating in—whether you are a Christian wanting guidance in the project of faith seeking understanding, or a non-Christian wanting to observe a Christian's pursuit of wholeness in our fragmented world.

Nicholas Wolterstorff
Professor of Philosophy
Calvin College &
The Free University of Amsterdam

# PREFACE

It is my hope that this book will be helpful to present and future students—indeed, all those who are perplexed by the issues that bring biology and Christian thought into contact. I offer here a book that speaks to these issues from the point of view of a practicing and deeply committed biologist—and I should add, a practicing and deeply committed Christian.

In the researching and writing process I have come to appreciate the great debt owed to those who have looked into the subject matters of this book and have put their thoughts in writing—especially those who have had a burden to bring knowledge from God's world into harmony with knowledge from God's Word. The back of the book lists many of those who have influenced my thinking. I also have a profound appreciation for the task of research in biology—for those who have been motivated to learn about how this world of life is structured and have chronicled their work in the journals. This is a part of biology that I have participated in and would not have missed for the world!

The Biology Task Force identified in the front of the book was thoroughly involved with the planning, writing, and revising phases and was a great support in what originally seemed to me an overwhelming assignment. I am also grateful to the members of the Series Advisory Board and to former Coalition President John Dellenback for their support and encouragement. Support from Gordon College in the form of a sabbatical and appointment as Staley Scholar has been crucial to this project. I would also like to thank my biology colleagues at Gordon—Jane Andrus, Tom Dent, and Russ Camp—who have sacrificed of their time in order to free me to complete this work. My wife, Ann, has been my first and most important reader and critic of each chapter, and I thank her for her patience and loving support.

Richard T. Wright
Wenham, Mass., 1991

# BIOLOGY THROUGH THE EYES OF FAITH

*Chapter 1*

# BIOLOGY AND WORLDVIEWS

You know, our fathers had plenty of deer and skins, our plains were full of deer, as also our woods, and of turkies, and our coves full of fish and fowl. But these English having gotten our land, they with scythes cut down the grass, and with axes fell the trees; their cows and horses eat the grass, and their hogs spoil our clam banks, and we shall all be starved.

MIANTONOMO, sachem of the Narragansett Indians, 1642

## A Meadowlark's Call

Joel Cooper was probably about sixteen when I met him; he seemed old and wise to my ten-year-old eyes. My family was spending three months in St. Petersburg, Florida, with my grandparents, while waiting for people to move out of a house we had purchased in New Jersey. Joel lived three houses down the street, and he was crippled—probably a victim of polio.

But Joel Cooper knew how to catch anoles (we called them chameleons) and striped skinks (another kind of lizard), and he knew where the cardinals nested and he had squirrels that would eat out of his hand. Joel showed me how to find the two owls—a screech owl and a barred owl—that roosted in the shelter of the Spanish moss that decorated a huge old live oak in his yard. I thought I was in paradise. School was something to endure. My life began at 2:30 in the afternoon, when I could explore Joel's world with him—the world of overgrown shrubbery and vines and citrus trees that sprawled behind his house.

For the first time in my life, I was seeing insects and lizards and snakes and birds, and I was amazed by what I saw. There was a world to be discovered out there, and Joel helped to peel the scales from my eyes so that I could—for hours on end—wander the orchards and palmettos and ponds and look for new things. My mother bought me the green and yellow and blue books of birds, and I began to sketch them from the pictures, wondering if I would ever get to see so many birds.

For some reason, I was particularly captivated by the picture of a meadowlark—a medium-sized bird with a brown back, and a black bib across a brilliant yellow breast. One day my family and I went to a potluck dinner following morning church. Naturally, being ten years old, I slipped away from the proceedings to see what I could find outside. Off in the distance I heard a clear and loud two-syllable bird song. Using all the stealth I could muster, I approached the source of the song, coming closer and closer and still seeing nothing. I was lying in the grass when a bird moved into a little sandy patch about ten feet away. To my great wonder and delight, it had a black bib across a bright yellow breast and sang the song I had been hearing—a meadowlark!

The memory of that instant has a permanent place in my mind's picture gallery. Since then, many other pictures have been hung there—a peregrine falcon darting into a flock of shorebirds, a European ruff appearing in a little pond in south Jersey, an immense flock of sandhill cranes soaring overhead in an Indiana state park, to name only a few. But that first picture of a meadowlark remains as a symbol of the joy and amazement that came when a new world began to be opened up to me. I hope that I told Joel about the meadowlark; it would have made him happy.

## The Study of Biology

There is much more to biology than watching birds, of course. Yet I wish that everyone who takes a course in biology would get to see their equivalent of my first meadowlark—something from the living world that simply captures the imagination in such a way that there is no turning back. From that point on, whether as an object of study or of simple observation, the living world opens up an unending array of complexity and the unknown, and the observer is transformed. Biology—better yet, the living world—has the capacity to amaze and delight, and to some it calls for a deep and possibly lifetime commitment.

As a former student and a teacher, I am aware of the frustration of having to study the world of biology as it is presented from the captivity of a course in college. If you are like many college students, a course in biology represents your only and last college contact with science as a system of knowledge. If someone asks you ten years from now what you remember from your biology course, your answer may be some unpleasant recollection of the laboratory, or at best an impression of how excited your professor was about the subject. Yet during that ten years, your life will certainly be profoundly affected by issues and realities of the living world.

How will this living world affect you? As a start, remember that *you* are a living organism. Your food, the very air you breathe, the water you drink, your wastes, your sexuality and reproductive activities, your sicknesses, the deaths of friends and relatives—these are reminders of your basic biological dependencies. Then consider that you are part of the largest human population ever to inhabit the earth, with an impact that far surpasses that of any previous human population or any other biological species. Your environment—not the city or town, highway or airport, but the living fabric—will be showing greater signs of stress than ever before because of human impacts on it. Then there is the world of ideas, opinions, and more important, faith. How should we help whole nations that are caught in a downward spiral of population growth and poverty and hunger? Shall our legal options include the right to abortion? Where, after all, do we come from? How shall we live?

These questions matter; and because biology is what we call

the system of thought and knowledge that deals with the living world, biology should also matter. As it is presented in most texts, biology is—or pretends to be—value free; it does not appear to render moral judgments concerning the issues that are raised as a consequence of our increasing knowledge of how living systems work. Yet biology is a thoroughly human enterprise, and a textbook will necessarily reflect the perspectives—and the worldview—of the authors. So, if you are approaching biology expecting either a neutral treatment of the subject or definitive answers to perplexing issues, you will be disappointed.

This book also reflects a particular worldview, but at least you will know what that worldview is. Beyond the problem of neutrality, however, I hope to convince you that biology is not only loaded with the potential to amaze and wonder, but it also bears a very large burden of important, value-laden issues that penetrate to the very depths of human existence. The issues are as controversial as they are complex. They range from fundamental questions about the real nature of science through concerns about origins; from medical and genetic knowledge to questions about how to live in the environment; and what constitutes justice in the distribution of basic human needs. The issues also make it clear that our scientifically derived knowledge has brought us great powers—and with those powers come equally great responsibilities.

This book is also meant to be a supplement to the information found in a typical biology text—not primarily to correct that information, although some correction may be in order; but to make every effort to view biology through the eyes of Christian faith. This perspective makes it possible—indeed, necessary—to deal with the issues brought to us by biology. It gives us a basis for making moral judgments, sometimes specific, but more often in the form of general ethical principles. In the moral wasteland that marks the end of the twentieth century, ethical principles and moral judgments are desperately needed.

More fundamental than ethics and morality is the Christian insight that to study biology is to study an important part of God's creation. Life in all of its manifestations—bacteria, birds, butterflies, even viruses—spreads before our eyes an array of creativity that has drawn people to faith for centuries. Many share a heartfelt conviction that this cosmos, this world of living, throbbing creation could not have originated in any other way than through the purposeful work of a powerful Creator. And indeed, the Bible tells us that this conviction is valid, and that the Creator's name is God.

On the other hand, the story told in biology texts seems to be something quite different: that life originated spontaneously, and that the array of living things, including humans, evolved over time as the result of undirected natural selection. The controversy over the question of origins has polarized scientists and many people of religious faith. Students are understandably perplexed by this issue, and Christians and non-Christians alike wonder if it is possible to be a "good" scientist and also believe the Bible—"What's a nice Christian girl like you doing in a white lab coat?"

In this book I will explore in some depth the biblical message of creation and relate it to our current understanding of origins. I will also bring out the biblical message of dominion and show how it applies to the interactions of the life sciences with society in medicine, genetics, and environmental concerns. In the process, I hope to show not only that it is possible to be both scientist and Christian, but that this combination is in fact the most rewarding and exciting way to approach biology.

### Worldviews

The issues and perspectives presented in this text are vitally connected to matters of belief, ethics, and relationships. A very useful approach to exploring these matters is to investigate worldviews. In this chapter, we will define worldviews and then begin to consider what constitutes a Christian worldview. The theme of worldviews will appear often in the text, and especially in the final chapter. To illustrate the vital importance of worldviews, let's go back in time to a crucial period for the landscape of New England: the years 1600 to 1700.

### *Worldviews in Conflict*

The Native American population of New England in the year 1600 was probably between 70,000 and 100,000, very similar to the size of the population of European settlers who occupied much of the same land 100 years later. These two cultures, however, interacted with the environment in profoundly different ways, and affected the landscape in ways that directly reflected the major worldview differences between them. It is a matter of history that these two cultures failed to develop a compatible relationship with each other; it is less well known that a major reason for this was the fundamentally different relationships of these cultures with the land.

The pre-Colonial landscape of the native tribes was probably a patchwork of woodlands, marshes, tree swamps, and occasional agricultural clearings. The woodlands were often relatively free from underbrush, the results of deliberate burning. The early visitors, accustomed to the European landscape, were impressed by the number and size of the trees, and particularly the abundance of fish and game. They failed to realize that this richness reflected a stable equilibrium that was the result of Native American culture or worldview. Historian William Cronon describes three elements that contributed most to their land tenure: their nomadic habitation, their notions of land "ownership," and their approach to "property."

The New England tribes (among them the Agawams, Narragansetts, and Wampanoags) moved as their food, shelter, and firewood needs changed. These needs, and the relative food resources, varied with the seasons. Spring found the tribes moving from the forests to the coast, where they could begin planting and also have access to the abundant migratory fish runs. In the fall, the coast also provided migratory birds and shellfish. They moved back into the woodlands and forests in late fall, where their shelter and firewood needs were better met. The pattern of life evolved around the seasons and was adjusted to the ecologically diverse landscape that provided them with food, game, firewood, and shelter.

Because they moved about with the seasons, the tribes seldom identified themselves with fixed locations. They did not conceive of owning land; land was useful only as it provided for their needs, and those needs changed with time. Tribes might maintain a kind of sovereignty over some areas, such as the woodlands where winter villages were constructed, but coastal resources were open for the taking. This sovereignty was limited to the right to take their living from the land.

The Native American notion of property was clearly tied to their nomadic tracking of the seasons. They had few "possessions": a few tools, weapons, clothing as needed. These were either light and mobile, or readily abandoned and refashioned when need arose. The idea of accumulating possessions so as to achieve a measure of wealth was absent from their culture. They accumulated little in the way of worldly goods because they required little. Their dwellings were built and torn down rather easily, clearly an adaptation to their nomadic life. To the European colonists, the Native Americans seemed strange: a people living in abject poverty, but surrounded by great natural wealth.

The European settlers came to New England as an alien culture. They were unable to comprehend tribal life and land tenure, and their worldview was shaped by the life they had known in Europe. Their habitation was governed by the European notion of property rights: Land was to be developed into villages and towns for permanent occupation. Native use of the New England ecosystems for harvesting food was not considered a legitimate claim to land: If the tribes did not actively occupy the land, they had no property rights to it. If the colonists "bought" land from the tribes, the parties to such a transaction had very different ideas about the rights conveyed (the Native Americans did not imagine that these were permanent transactions preventing them from further access to the land).

Another element in the European worldview was the identification of forests, fur-bearing animals, and fish as "natural resources"—not simply to provide for their needs, but to be harvested in order to generate wealth. Very quickly, the New England colonies were linked to the European economy by a lively trade in fur, fish, and timber. The new land—abundant in wildlife and forests when the colonists arrived—was thus defined in economic terms. Land was a commodity, to be bought and sold; the landscape was shaped by European-style agriculture and by the domestic animals brought over by the settlers; and the abundance of wildlife and timber was harvested and turned into wealth.

The landscape very quickly began to reflect these new relationships: The beaver population was soon decimated; there was no significant fur trade by 1700. Turkeys and deer disappeared from the coastal areas of New England. The timber harvest moved further and further into the wilderness, and very soon there were firewood shortages in the coastal New England cities and towns. The land lost much of its agricultural productivity as a result of overgrazing and poor farming practices. As a result, the land was rendered uninhabitable for the native tribes; it no longer supported the abundance of game, and was bounded by fences and protected by property rights.

The translation of the New England landscape into economic production was ecologically self-destructive. As Cronon points out, the colonists did not learn the difference between yield and loot, and the results were very soon evident.

There are lessons in this history that clearly go beyond simply illustrating the differences between worldviews and their consequences, and I would be remiss not to mention them in passing: (1) Natural ecosystems are capable of supporting human life on a sustainable basis, but will do so only if ecological realities are recognized and respected. (2) Economic and political power determine the fate of much of the natural environment, especially when the elements of the natural landscape are defined in economic terms. (3) Misuse of the land and resources can result in permanent changes. It is probably fair to say that our land use still demonstrates its European origins, and that we still have not learned well the lessons so sadly and thoroughly illustrated by the land tenure of the Native Americans and the colonists.

### Worldview Defined

We have just considered an illustration of two worldviews, two different approaches to the totality of life experienced by two cultures. In this case, one worldview swamped the other, and the consequences for the less powerful society and for the landscape were profound.

Worldviews are vitally important. James Sire, whose book *The Universe Next Door* is rich source material for a study of worldviews, defines a worldview as "a set of presuppositions (or assumptions) which we hold (consciously or subconsciously) about the basic makeup of our world." Clearly, everyone has a worldview. Worldviews determine values, help us to interpret the world around us, and in general function as a guide to life. Another definition of a worldview is "the comprehensive framework of one's basic belief about things." A worldview is a kind of picture of how we think the world to be, a picture that can often only be seen by observing how a person lives. The reason for this is that few people ever seriously analyze their worldviews. Most would have difficulty in laying out the "framework of their basic beliefs." Yet everyone has a set of basic beliefs, convictions that usually hold together in a pattern that some refer to as a "system of values."

As a guide to life, a worldview functions in the same way a compass and a navigational chart might be used to steer a boat through coastal waters. Our worldview helps us to interpret what we see, the way we evaluate the events that cross our pathway each day. It also guides us in determining direction, in making choices. In this sense, then, we would say that a worldview is normative—it deals with our sense of what is right, or good, or necessary, or possible. As creatures who must continually make conscious decisions, we need the kind of guidance and perspective that a worldview provides.

It is also important to note that our actions do not always accurately reflect our worldview. Our behavior and our beliefs are often inconsistent. However, we are aware of the inconsistency, and our basic charted course is maintained, even though we may temporarily deviate from the course. Of course, it is possible that our worldview also includes a great willingness to forgive these little inconsistencies!

Where do we acquire our worldview? Although the sources of a worldview are many, we usually acquire the fundamental components from the culture in which we live. Because of this, worldviews are shared phenomena. They help to define the uniqueness of a society, they provide much of what binds people together into a community. A culture provides a vision for life in all of the dimensions in which we function: work, education, politics, recreation, art, family, neighborhood, legal concerns, financial affairs. And there is continual feedback between the way in which a culture shapes the worldview and the way in which the communally held worldview continues to shape the culture.

## A Faith Commitment

Guidance, perspective, basic beliefs, system of values—these terms suggest to us that a worldview has many of the characteristics we normally associate with religious belief. In their challenging book *The Transforming Vision*, Brian J. Walsh and J. Richard Middleton describe four basic questions everyone faces, questions that are foundational to living out our lives: (1) Who am I? (What are human beings and what is the purpose of human existence?) (2) Where am I? (What kind of world and environment do I live in?) (3) What's wrong? (What hinders me from fulfilling all of the basic purposes of my existence?) (4) What is the remedy? (How can I overcome the obstacles to achieving this fulfillment?) It is clearly impossible to answer these questions without probing our fundamental beliefs.

Even if we never examine our beliefs, they still exist and influence our approach to life. For example, if I believe that my first priority is to look out for Number One (question 1), and that the world is basically unfair (question 2), and that I don't have my share of the good life (question 3), then I will tend to turn all situations to my advantage (question 4), even if I have to bend the rules—but I must be careful not to get caught! On probing more deeply into the reasons for these beliefs, I would probably admit that I don't really believe in a God who cares about people, and that science has proven that God doesn't exist anyway, and that my actions therefore are curbed only by a vague sense that I shouldn't hurt other people. Although not at all admirable, this is the essence of a worldview.

## A Christian Worldview

Christians should have worldviews that are strongly related to their faith. However, some Christians hold worldviews inconsistent with biblical truth and often heavily influenced by surrounding culture. There is a strong, consistent temptation to adopt elements of a cultural worldview, a worldview that may have little regard for Christian truth. The Christian's task, then, is to shape a worldview according to the teachings of

Scripture, and continually test these worldview beliefs against the Scriptures. A biblical worldview should then serve as a guide through life, and this includes responses to questions of origins, stewardship, justice, medical and genetic concerns—all issues that will be raised in this book.

A worldview begins with presuppositions, fundamental beliefs that pave the way for what follows. A most important presupposition of a Christian worldview is that the Bible is God's revelation to humanity; its purpose (2 Tim. 3:16–17) is to teach, rebuke, correct, and train us in righteous living so that we may become people of God prepared for a life of service. This means that Christians are to take biblical teaching seriously, to let it inform, shape, and correct their worldview. Since a worldview involves our approach to all aspects of life and culture, we can expect Scripture to speak to the totality of life—not simply some sector that we might be tempted to mark off as being "religious" (as opposed to "secular").

Beyond this presupposition lies the wealth of doctrine concerning faith and practice that has come to be identified with orthodox Christianity. In my view, there is a particular stream of thought within Christianity that expresses well the fact that a biblical faith embraces all of life (a truly *world* view). One summary of that approach has been given by theologian Herman Bavinck, as follows: "God the Father has reconciled his created but fallen world through the death of His Son, and renews it into a Kingdom of God by His Spirit." A shorter way of expressing this view is to say that "*grace restores nature*," meaning that the work of Christ in redemption brings about the restoration of God's original good creation. This biblical worldview provides a framework for life, but does it also provide an approach to organized disciplines like biology, an approach that will make possible a Christian interpretation of that discipline?

In *The Universe Next Door*, James Sire shows that although there seems to be almost as many worldviews as there are people, most are shown to be variations on a limited number of themes. Sire catalogues a number of these dominant worldviews that are nontheistic—that is, they assume that the God portrayed in the Bible does not exist. One of the worldviews—naturalism—represents the belief that "The Cosmos is all that is or ever was or ever will be," as expressed in the opening words of Carl Sagan's popular television series and book, *Cosmos*. Naturalism rejects the concept of a supreme being and sees humans as nothing more than the product of an evolutionary process. Biology—as it deals with the origins of life, of the vast array of living organisms, and of humankind—is often presented in a thoroughly naturalistic framework. On the other hand, the biblical worldview ascribes these origins to the creative activity of God, and maintains that the created order is dependent on God. Here two worldviews are clearly in conflict; each is exclusive of the other.

As the book progresses, you will encounter a number of worldview beliefs that claim the allegiance of many and often represent the dominant view of our culture: naturalism, evolutionism, humanism, secularism, scientism, materialism, economism—to name a few. It is vitally important that you see these for what they are: faith systems in conflict with biblical Christianity. Biology—in its compilation of information about the world of life—presents us with many opportunities to probe human thought and behavior from the perspective of Christian thought.

## Summary

The living world has a remarkable capacity to amaze and delight us, often taking us by surprise when it does. Once our attention is caught—whether from firsthand experience or in a biology course—we find that biology deals with many value-laden issues that profoundly affect human life. In particular, the issue of how life originated continues to concern people. This book explores in depth the themes of creation and dominion, two biblical themes that inform the study of biology.

The perspective of worldviews is introduced by examining the impact of two cultures on the landscape of New England: the original tribal inhabitants, and the European colonial settlers. From 1600 to 1700, the New England landscape changed from a rich and productive natural mixture of habitats that sustained the native tribes, to a heavily exploited land that lost much of its richness and basically reflected the European landscape. The two worldviews were incompatible, and the land, animal, and Native American populations suffered as a consequence.

A worldview is defined as a guide to life, a basic set of values that we acquire primarily from our culture. At the foundation of a worldview is a commitment to beliefs about our existence, our basic problems, and how we solve them—a faith commitment. For Christians, the Bible provides a worldview that serves as a guide to all of life—if it is consistently followed. This text will explore biological science from the perspective of a biblical worldview, and will examine other worldview beliefs that clearly influence the development and interpretation of scientific knowledge.

*Chapter 2*

# GOD AND HIS WORLD

By faith we understand that the universe was formed at God's command, so that what is seen was not made out of what was visible.

HEBREWS 11:3, NIV

## Perspectives: The Biblical Picture

Biology does not ordinarily direct students to the Bible for information about the subject. But to see biology through the eyes of faith, we need to get God's view of things early in the process. One strategy is to approach the subject with the basic question: *What does Scripture say about the relationship between God and the natural world?* As we answer this question, we will hopefully lay the foundation for a biblical worldview that supports our efforts to develop a Christian perspective on biology.

### *God and Nature*

The word "nature" does not appear in the Old Testament. The Hebrew writers did not have a word that would translate into what we commonly understand as nature (the material world as an independent reality); it was a concept foreign to their way of thinking. The Greeks had a word, *cosmos*, which New Testament writers used to indicate the whole created order, as in the first chapter of John's gospel (John 1:10). A preferable term for "nature" (cosmos in modern times has become synonymous with nature) is "the creation," and there are several reasons why this distinction is important. "Nature" commonly stands for all matter and energy—everything from the universe to earth to the living world. We read that nature is governed by "natural laws," and these operate on the basis of cause and effect to bring about all that happens in this "natural" realm. In this view, *Nature* (with a capital N), is autonomous—it has independent status. The *Creation* (again using capitals to indicate a special use) also refers to the entire material world, but the very term we use indicates its status: The Creation is dependent on God for existence. This is a crucial difference.

The distinction between Nature and Creation clearly has worldview implications. Nature as we have defined it is a fundamental part of the worldview identified as *naturalism,* which holds that there is nothing beyond the material world. The significant faith component of this worldview is suggested by our common use of the term "Mother Nature," as we ascribe to this autonomous world the powers of creation (for example, "Mother Nature has provided the arctic fox with its white coat").

The Christian counterpart to this worldview is called *theism.* Theism holds that God is the central fact of existence, and that the material world exists only in dependence on his creative and sustaining activity. From this perspective, there is no possibility of an autonomous, self-sufficient Nature. Because of these considerations, it would be more appropriate to use the term Creation as a substitute for Nature when we refer to God's created world. However, the term Creation is so thoroughly identified with one particular view of how the world was made, and the term Nature is so commonly used to refer to the natural world (often without worldview implication), that to carry out this substitution would cause confusion. Let us use the terms nature and creation interchangably, and simply be aware of this distinction as we do. We will now turn to Scripture to explore some themes concerning God and his creation.

## Creation: God Brings All Things into Being

The writer of Genesis opens the Bible with magnificent, sweeping grandeur: "In the beginning, God created the heavens and the earth" (Gen. 1:1). God's existence is simply assumed, and his transcendence is declared in this opening statement. It is a statement made to a people—Israel—who already knew God as their redeemer. The order is important; faith in God as redeemer comes before faith in God as creator. The fundamental purpose of this part of Scripture, then, is to inform God's people more fully of who their God is. Knowing this, they can worship and praise God more completely, and the Psalms in particular demonstrate this principle.

As Genesis develops, we read of God calling the various facets of creation into being. In a later chapter, we will investigate the details of the description of God's acts of creation as we try to understand the reasons for the continuing controversy over origins. Our present focus is more broad: What is the tesimony of all of Scripture regarding the creation? The clear message of Scripture is that God is responsible for bringing into existence the universe and all that it contains. But we are curious creatures—we would also like to know how he did it!

### *Creation by Word and Wisdom*

One image used throughout the Bible to describe how God creates is the image of creation by his *Word*. Early Genesis pictures God speaking, and the Creation coming into being in response: "And God said . . . and it was so" (Gen. 1). The Psalms give testimony to creation by the Word: "By the word of the Lord were the heavens made, their starry host by the breath of his mouth. . . . Let all the earth fear the Lord; let all the people of the world revere him. For he spoke, and it came to be; he commanded, and it stood firm" (Psalm

33:6,8,9). We read in 2 Peter 2:5, that "long ago by God's word the heavens existed and the earth was formed . . ." Although the use of this image certainly presents God's majesty and power, its primary orientation is to God's *authority*. His authority is absolute; he speaks, and creation obeys.

The other primary image used to describe God's creative work is his creation by *Wisdom*. The primary orientation of this image is toward *purpose*—the fulfillment of God's intentions for the development of the created order. Proverbs 8:22–31 speaks eloquently as "wisdom":

> The Lord possessed me at the beginning of his work, before his deeds of old; I was appointed from eternity, from the beginning, before the world began. . . . I was there when he set the heavens in place, when he marked out the horizon on the face of the deep, when he established the clouds above and fixed securely the fountains of the deep, when he gave the sea its boundary so the waters would not overstep his command, and when he marked out the foundations of the earth. Then I was the craftsman at his side. I was filled with delight day after day, rejoicing always in his presence, rejoicing in his whole world and delighting in mankind.

### *The Creation of Life*

Nothing in creation is more majestic or more symbolically rich than the creation of life by the living God. Genesis 1:11–26 pictures God populating domains of earth, air, and water by living things, culminating in the creation of humankind. In doing so, God has given to his creation a power held only by him, and therefore only his to dispense—the power to create. This is an essential characteristic of living things, one that clearly separates life from non-life. In a real sense, this ability of living things to reproduce is a reflection, however dim, of God's power to create. It is a limited power ("according to its kind"), but a very special one, one that pleases God. He not only declares his created life forms good, he adds his blessing to this unique part of the creation, encouraging life to exercise its potential and reproduce, so that crawling and swimming and flying creatures can occupy the earth, air, and water (Gen. 1:22). Speaking of animal life, the psalmist comments on the intimate relationship between God and life: "When you take away their breath, they die and return to the dust. When you send your spirit, they are created, and you renew the face of the earth" (Psalm 104:29,30). As you learn about life, consider how the creative abilities of living things reflect God's own power to create.

### *Creation's Source*

Where did matter and energy originate? The answer is a matter of faith and a clear indication of the crucial role of world views. G. G. Simpson, Harvard paleontologist and author of numerous books on evolution, states that

> the origin of the cosmos and the causal principles of its history remain unexplained and inaccessible to science. Here is hidden the First Cause sought by theology and philosophy. The First Cause is not known and I suspect that it never will be known to living man. We may, if we are so inclined, worship it in our own ways, but we certainly do not comprehend it.

The biblical worldview identifies the First Cause as God. The writers of Scripture affirm that the universe and all that it contains came from God, who created (and continues to create) by his Word and by Wisdom. If we need to postulate an ultimate source, we simply point to God. There is deep mystery here, perhaps never to be fathomed by human reasoning. God does not tell us, and perhaps we wouldn't understand him if he did. Yet creation is an expression of God's being; and although we may not understand how it came into being, we can perceive something of his immense power and wisdom as we look at what he has done.

### *Creation's Status*

How should we consider the cosmos—the world of matter and life? Is it really autonomous, independent of any external reality as affirmed in the naturalistic worldview? Or is it to be considered as containing godlike qualities, either as seen in various forms of paganism or in more modern versions of humanism? These are significant questions, for the answers will both reveal and determine the worldview of the questioner. What is, in fact, the status of the cosmos, of nature? The Bible speaks clearly: It is God's creation. He is responsible for its origin and, as we will see, for its continued existence. This is the most important message of early Genesis, and the rest of Scripture gives testimony to this message. But it takes the eyes of faith to see it.

## Governance: God Sustains His World

### *Natural Laws and Creational Laws*

One of the fundamental assumptions made by scientists, and indeed by almost everyone, is that matter and energy behave in an orderly fashion. When we bend our minds to the study of matter and energy we see regular patterns of behavior, so much so that we are confident that we can predict that behavior for other times and places. Scientists routinely refer to these regular patterns as the *natural laws;* many believe that these laws actually govern nature—that nature is autonomous. But consider the cause-and-effect relationships behind snow, hail, frost, and wind. This seems to be a matter well taken care of by natural laws. Yet Scripture testifies that these are God's work: "He sends his command to the earth; his word runs swiftly. He spreads the snow like wool and scatters the frost like ashes. He hurls down his hail like pebbles. Who can withstand his icy blast? He sends his word and melts them; he stirs up his breezes, and the waters flow" (Psalm 147:15–18).

The point is, there is a critical distinction between a scientific explanation of the way that natural systems function, and the actual forces and relationships that are being described. A "natural law" is a description, a model. It is a human product, at best a tentative approximation of reality. For example, in the governance of a society, a law is a description of the expected behavior of people or corporations; however, the actual governance is in the hands of the appointed or elected political authorities who see to it that the law is obeyed. So with the natural world: We may decide to call our models of the way matter and energy behave laws, but we only deceive ourselves if we think that these laws are actually the power behind the universe. We need to look beyond science to delve into such matters.

In fact, the lawful response of the creation to God may be described as the outcome of *Creational Law*. This law has two modes of operation: structural law and normative law. In *structural law*, God establishes and sustains the cosmos, upholding it by his power. As we will see below, this is a law-bound covenant relationship between God and the natural world, producing the order and regularity that our observations perceive. Our attempts to describe the outcome of God's structural law are legitimate scientific activity, but the resultant theories and laws are at best tentative shadows of the real thing.

*Normative law* refers to the proper form and function to be taken by the created order as it develops over time. The distinction between these two forms of creational law is most critical in the case of humanity. In our case (as opposed to other life forms), God has given us his written, normative law. He has created us in his image, and he commands us to live in obedience to him and to worship him for who he is. We bring glory to him when we do. As we well know, such norms can be violated, and God's written word makes it clear that we will be held accountable for violating his normative laws.

I believe that God's normative law also applies to other life forms, in the sense that they have developed over time and display ecological relationships with each other and with the physical world. Their existence and interactions show God's glory and wisdom as much as do the more structurally law-bound elements of the creation (the heavenly bodies, for example), as the Psalms testify. These normative relationships can be broken, however, as the result of human activity. Species are brought to extinction, and natural ecosystems are degraded; the clear result of this is that God's glory is no longer displayed. But we digress.

### *God's Covenant with Creation*

God's relationship with his creation actually has the status of a covenant, an agreement as made between sovereign and subjects. God speaks through the prophet Jeremiah: "If you can break my covenant with the day and my covenant with the night, so that day and night no longer come at their appointed time, then my covenant with David my servant . . . can be broken . . ." (Jer. 33:20,21). The order and regularity that we observe in the natural world is a matter of obedience to what we have called Creational Law, where use of the word law also includes the element of command. It is important that there is the expectation of faithful obedience on the part of the creation, and faithful rule on God's part; the covenant is kept.

This covenant relationship is the meaning behind the Christian claim that God sustains his world. He is the reality behind matter and energy; these properties of the natural world would not exist except for God's continuing activity in faithfulness to his covenant. Psalm 104:13–17 is an eloquent statement of God sustaining the different parts of his world:

> He waters the mountains from his upper chambers; the earth is satisfied by the fruit of his work. He makes grass grow for the cattle, and plants for man to cultivate—bringing forth food from the earth

. . . The trees of the Lord are well watered, the cedars of Lebanon that he planted. There the birds make their nests; the stork has its home in the pine trees . . .

## The Value and Purpose of the Creation

### *Creation's Goodness*

Genesis 1 testifies repeatedly to the goodness of the Creation. It is important to note that in declaring creation good, God was making a statement about its value. Good means having inherent worth; value is intrinsic—it is built into the things God created. In declaring creation good, God was also declaring his pleasure with all of his creative work. Why was he pleased? Possibly because he saw the unfolding of creation as an obedient response to his Word, one worthy of the covenant between God and his creation. Also, he might well have been pleased because what he saw was an expression of his wisdom; God recognized that the creation was majestic enough to show his glory. In other words, the goodness of the creation reflected God's own goodness. But does it still do so? After all, sin has entered the world. Can a fallen world still be good?

In chapter 9 we will consider in detail the Fall of humankind and its effects on the rest of the creation; however, some comment on one impact of the Fall is appropriate. Some writers have speculated that when God warned Adam and Eve that death would come of eating from the tree in the middle of the Garden of Eden, he meant biological death in general. However, we must assume that Adam and Eve ate plants, if not animals ("You are free to eat from any tree in the garden" [Gen. 1:16]), and therefore the plant tissue had to die. And since animals must also eat other animals and plants in order to "be fruitful and increase in number," as they were commanded to do from the beginning, we may conclude that biological death is as old as life itself. Psalm 104:21 speaks of the lions seeking their food from God, evidence that death after the Fall is still part of God's plan. Indeed, the insight from a study of ecology reveals that only death makes more life possible. May we not conclude that biological death is part of God's providential care for his creation?

Is the creation still good? The Scriptures give us a clear answer: Creations's goodness has persisted and continues to the present. "The heavens declare the glory of God; the skies proclaim the work of his hands. Day after day they pour forth speech; night after night they display knowledge. There is no speech or language where their voice is not heard" (Psalm 19:1–3). In Acts 14:15–17, Paul appeals to the crowd in Lystra to turn

> to the living God, who made heaven and earth and sea and everything in them. In the past, he let all nations go their own way. Yet he has not left himself without testimony: he has shown kindness by giving you rain from heaven and crops in their seasons; he provides you with plenty of food and fills your hearts with joy.

And in Romans 1:20, we read that God's creation has testified so plainly to him that people should have known and worshiped him instead of the creation: "For since the creation of the world God's invisible qualities—his eternal power and divine nature—have been clearly seen, being understood from what has been made, so that men are without excuse."

Paul continues in that passage to address the folly of making idols of the creation: "images made to look like mortal man and birds and animals and reptiles" (Romans 1:23). It has been a human tendency over the ages to idolize and worship nature. In a sense, this worship is understandable. It is, at least in part, a response to the beauty and grandeur of the creation, and a recognition of our dependence on it. It is also, however, often a sign of fear, a response to the unknown and the power displayed by the natural world. Scripture makes very clear that those qualities should direct our attention to the Creator. It is ironic that the very goodness of the creation, which is just a reflection of God's goodness, diverts our attention away from the God whose power created and sustains it. The continuing goodness of creation is there for us to see. People, mushrooms, salt marshes, forests, mountains, stars—all creatures great and small—proclaim the power, wisdom, and glory of God.

### *"Praise Him All Creatures Here Below"*

Many passages from the Psalms support this second line of the Doxology. Perhaps Psalm 148 expresses this to the fullest:

> Praise the Lord from the earth, you great sea creatures and all ocean depths, lightning and hail, snow and clouds, stormy winds that do his bidding, you mountains and all hills, fruit trees and all cedars, wild animals and all cattle, small creatures and flying birds, kings of the earth and all nations, you princes and all rulers on earth, young men and maidens, old men and children. Let them praise the name of the Lord, for his name alone is exalted; his splendor is above the earth and the heavens. (Psalm 148:7–13)

Since all but humankind lack articulate speech (as far as we know!), we may assume that the rest of the creation praises God by its very existence and activity. This is the highest purpose of the creation: to bring glory to God. Our responsibility as creatures in his image is not only to praise him, but also to recognize that the rest of the creation is praising him. As students of biology, we should be quick to turn our attention from the object of our study to the God who created and sustains his creation, and be able to see that the living world does indeed bring him glory. "Let everything that has breath praise the Lord" (Psalm 150:6).

Why should all creation praise God? Again, the testimony of the psalmists brings us the answers: for his wisdom (Psalm 104:24); for his unfailing love (Psalm 107:21); for the faithfulness and justness of his works (Psalm 111:7), for his glory

(Psalm 113:4); for his unfathomable greatness, his mighty acts, his majesty, his righteousness (Psalm 145:3–7).

In spite of the clear testimony of Scripture that the primary purpose of the creation is to bring glory to God, there is a prevailing attitude, even among Christians, that the only purpose of the creation is to satisfy our needs and wants. The flip side of this attitude is to question the value of anything that apparently serves no useful purpose for humanity (what good are slugs, or slime molds, or mosquitos, or swamps?). This is a potentially destructive attitude, for it allows us to treat nature as if its only purpose was to fuel the engines of progress—a completely utilitarian approach that translates nature into human resources. Not only is this attitude indefensible from a theological point of view, it is clearly part of the network of causes for the environmental problems that are plaguing us. We will deal in depth with these issues later.

### *To Know the Creator*

The Scriptures proclaim that creation declares its maker, and that the testimony of the creation is so strong that we are without excuse if we do not see in it the evidence of a creator-God. Before the coming of Christ, this was God's only testimony to the gentile world, unless they came in contact with his chosen people Israel. However, when Christ came, announcing the Kingdom of God and bringing redemption by his death on the cross, the Apostles were led to record those events for all to read and respond. God has now become more fully known through his written Word, although we are not to think that creation has stopped testifying of him (Rom. 1:20).

God's purposes in creation and in redeeming his creation are now out on the table. In fact, Scripture testifies that Jesus Christ is not only the redeemer-God (John 3:16) but also the creator-God. John's gospel, speaking of Christ: "In the beginning was the Word, and the Word was with God, and the Word was God. He was with God in the beginning. Through him all things were made; without him nothing was made that has been made" (John 1:1–3). In Colossians 1:15–17 we read: "He is the image of the invisible God, the firstborn over all creation. For by him all things were created: things in heaven and on earth. . . . all things were created by him and for him. He is before all things, and in him all things hold together."

This then is the greatest purpose of the creation from our perspective, its greatest utility to us. If we follow our observations of beauty, order, and purpose in the natural world to a realization that all of this speaks of a Creator, God invites us to take the next step and find that the Creator has also come to redeem his creation. That redemption came at the cross, and if the creation can bring us to the cross, then we can know our Redeemer through faith. This opens up a new life, with new responsibilities but new power to carry them out. One of these new responsibilities is to be a partner with Christ in redeeming his creation, but we will wait for the last chapter to look into this matter.

### *Know the Creation*

Biology is much more than preserved frogs and fetal pigs. These common laboratory specimens are a poor reflection of the creation. As a student of biology who might want to see in your studies much more of God's glory, you need to leave the laboratory and get out into the natural world. It is no accident that David and other psalmists could write so beautifully of the creation—they were there, tending sheep or crops and living much closer to the real world than we do. Our automobiles, buildings, and shopping malls wall us in and dull our senses until we come to believe that they are the real beauty to be seen from day to day.

Walk in the woods on an autumn day and listen to the leaves fall, and think about how important those leaves are to life. Buy a field guide to ferns, birds, rocks, butterflies—anything—and learn to identify and understand some of the diversity and beauty to be found in all corners of the natural world. When you do, teach what you have learned to children who are still young enough to love learning about such things. Invest some small part of your time in college at Au Sable Institute in the north woods of Michigan, where Christian stewardship of God's creation is the central mission. When you return to your studies—whether they be biological or economic—you will be better prepared to understand that at the root of our biological life or our economic activities is a great creation that is bursting to tell of the glory of God, and that to learn more about it is a high calling.

## SUMMARY

Biblical information about the relationship between God and his world is foundational to developing a Christian view of biology. Several terms are used to refer to the natural world: *nature, cosmos,* and the *creation*. A naturalistic worldview sees nature as autonomous, whereas a theistic worldview sees the natural world as creation, dependent on God. Although they have worldview implications, common usages suggest that these terms are interchangeable.

The Bible opens by declaring that God brought into existence the universe and all that it contains. Creation by God's *Word* is one image used throughout the Bible, and it indicates God's absolute *authority* over creation. Creation by *Wisdom*, another biblical image, speaks of *purpose*—the fulfillment of God's intentions towards the development of the created order. The creation of life seems to be a particularly appropriate creation by a living God, as his power to create is reflected in the creative powers of life forms. We are simply told that the ultimate source of the creation is God. He is responsible for its origin.

Scripture also testifies that God sustains his universe. The regularity and order that we observe in the natural world are described by science as *natural laws*, yet these are only models and not the true power behind the universe. The creation exists in a lawful response to God, in obedience to what is called *Creational Law*. This relationship has the status of a covenant; God faithfully upholds the creation by the word of his power, and it obeys God faithfully. Creational Law consists of *structural* and *normative* modes of operation.

Scripture declares the goodness of the creation; it has inherent value. Its goodness speaks of his goodness, and it brings glory to God. This ability to show God's glory has persisted, although since the Fall humankind has often worshiped some part of the creation rather than the creator. The primary purpose of the creation is to show God's glory, and Scriptures call on all creation to praise him. From a human perspective, the greatest purpose served by creation would be to lead us to the cross of Christ, where we can come to know him as both creator and redeemer.

*Chapter 3*

# THE SCIENTIFIC ENTERPRISE

The study of the basic philosophies or ideologies of scientists is very difficult because they are rarely articulated. They largely consist of silent assumptions that are taken so completely for granted that they are never mentioned. The historian of biology encounters some of his greatest difficulties when trying to ferret out such silent assumptions; and anyone who attempts to question these "eternal truths" encounters formidable resistance

ERNST MAYR

## About Science

This chapter is about natural science: what it is, how scientists think, the role of worldviews in scientific controversies, the place of biology in the sciences, and the limitations of science. In the previous chapter, we looked at the natural world as the creation, exploring the biblical statements on the relationship between God and his world. This chapter presents another approach to the natural world, a second way of looking at the creation. We will see that although these two approaches are different and distinct from each other, they are not necessarily in conflict, and for good reasons. To give us an entry into this rather formidable list of topics, let's take a look at some research that I have been doing in the salt marsh estuaries of northern Massachusetts.

### *An Affair with Bacteria*

O'Brien's Law states that inside every large problem is a small problem struggling to get out. The large problem, the overall goal of my research, is to achieve an understanding of the factors that control the free-living bacteria found in estuaries and coastal waters. To continue this research, I have to maintain a laboratory with a full-time assistant and some student help; this takes large sums of money, which I must obtain by way of grant proposals to the National Science Foundation. Once or twice a year, I publish a paper that deals with some aspect of my research; and I attend a meeting or two a year at which I get to present my current work.

In my large problem—the factors that control the free-living bacteria—two factors in particular have emerged as crucial: *substrate*, meaning the organic matter that nourishes the bacteria; and *grazing*, a function of the animals that feed on the bacteria. I have recently developed a mathematical model that shows how these two factors can interact to produce a given level of bacterial density (numbers). This model has proven to be useful in understanding the data we collect from the estuaries. It took me several years to understand the microbial food web well enough to develop the model, which is basically an adaptation of some equations developed more than forty years ago by a French scientist. The development of my research demonstrates the steady increase in knowledge and understanding that is part of the growth of a science and of a scientist. Once in a while, however, research takes an unexpected turn, as it did several summers ago when we were thinking about the problem of grazing. A fascinating smaller problem struggled its way to the surface and got out.

We wondered what happens to all of the bacteria that are continually feeding on the organic matter in salt-marsh estuaries. There are up to 10 million bacteria per milliliter, and they double their numbers daily in the summer. Our estuaries are heavily populated by three species of bivalve molluscs, which remain in place in mud or sediment and filter their food out of the water that passes by them. At the time, most people assumed that bacteria were too small for the bivalves' filtering apparatus—which does quite well on much larger particles like algae. But we were not aware that anyone had ever tested this assumption. So we brought a few ribbed mussels into the laboratory and exposed them to seawater containing natural bacteria. Then we took several samples over time, and with the use of a fluorescence microscope we counted the bacteria in the water. The results told us immediately that ribbed mussels do a rather good job of filtering bacteria from seawater. So we turned to the other two species of bivalves that are dominant in the estuaries, the blue mussel and the soft-shelled clam, fully expecting similar results. To our surprise, neither of these bivalves could filter bacteria. When we repeated and verified the results, we decided that we were on to something new.

Our next step was to test the problem in the field. Blue mussels live in dense beds (over 3,000 per square meter) in the lower intertidal waters of Essex Bay. We selected a bed that was about 50 meters wide, quite flat, and exposed at low tide. Catching an incoming tide just as it began to move across the bed, we sampled the water before and then after it passed over the 50 meters of mussel bed. We sampled every five minutes until the rising water level drove us into our boats, and then returned to the laboratory to analyze the water samples. The results showed that the mussel bed reduced the algae population in the water by 75 percent during one pass over the bed, but had no effect on the bacteria—excellent agreement with our previous findings. It was not possible to do a similar experiment with the ribbed mussels or soft-shelled clams, because they do not live in dense beds.

We discovered the final piece to the puzzle a few months later. We wondered if the differences in ability to filter particles were based on the spacing of the cilia on gill surfaces in the bivalves, since these animals are known to feed by passing currents of water over their gills. We measured this spacing in all three species and found that the cilia of the ribbed mussel are much more closely spaced, close enough to catch bacteria. The bacteria simply slip through the mesh in the gills of the other species.

These differences correlate very nicely with the ecological distribution of the bivalves. The ribbed mussels live in salt-marsh peat, farther upriver in these estuaries than the blue mussels. Bacterial densities are much higher in the water that circulates through the salt marshes each day than the water that bathes the blue mussels and soft-shelled clams closer to the ocean. We concluded that the bacteria are indeed an important source of food for ribbed mussels, and that these mussels are ideally adapted to exploit the high bacterial productivity of the salt marshes.

### *An Evaluation*

It should be evident from this account that we were not following some rigidly logical procedure, some well-defined set of rules, to get out this information about ribbed mussels and bacteria. We started with some broadly accepted knowledge: Bacterial production and numbers are high in salt-marsh estuaries; there are a lot of bivalves in these estuaries. Our search through the literature told us that nothing was known about the ability of the bivalves to feed on bacteria. We decided to try a simple experiment, and the results of that experiment led us to look further at other bivalve species, and to take the problem into the field. Our findings led us to being able to make a statement about ribbed mussels and bacteria, certainly not a major breakthrough but nevertheless a discovery, one that has subsequently been confirmed by other workers.

## How Scientists Think

In carrying out the research described above, we never questioned whether our procedure was in the right order, or thought about whether we were being properly objective or were using induction or deduction. I have a strong feeling that our statement about ribbed mussels and bacteria is correct; other research workers have accepted this concept and have put this information to use in related investigations. I believe that this is about as far as most scientists go in doing philosophical thinking about their work. That is not to deny that there is a very legitimate role for the philosophy of science; it's just that not many scientists bother to think about it. Yet with science—and biology in particular—playing such a prominent role in our society, how scientists think is vitally important. Even a very basic understanding of the philosophy of science would go a long way toward resolving many of the existing science-faith problems. Let's probe a little more deeply.

### *The Scientific Method*

Biology is the science of living things. As a natural science, it provides explanations for the observable array of organisms, and for processes and events involving those organisms. Most people simply assume that these explanations are correct. The knowledge in the biological texts is presented in a very positive light; it looks and sounds like the truth. Seldom, however, does a text clearly establish the criteria that were used to obtain the information in the first place and to select from that the material to be put in the text. To be sure, the general biology texts almost always present "the scientific method" in the first chapter, and there is a uniformity in these presentations that suggests broad agreement about this element of the philosophy of science.

In reality most texts are presenting a perception that is at least twenty years behind what the philosophers of science are saying. It is not that your textbook "scientific method" is all wrong; it is too simple. It fails to include some elements in the practice of science that are vitally important. So when you read about the "hypothetico-deductive method"—of formulating hypotheses, testing these hypotheses by forming predictions that lend themselves to controlled experiments, and establishing a theory based on the results, you should realize that this is an incomplete and somewhat out-of-date picture of how science is done.

## *The Components of Science*

The contemporary view of the philosophy of science, according to Calvin College philosopher Del Ratzsch, holds that there are three major components to the structure of a science: *data, theories*, and *shaping principles*. The *data* of biological science refer to the information that can be extracted in the form of measurements and observations concerned with living things and their surroundings. Ratzsch goes on to suggest that the gathering of these data must conform to one of the crucial touchstones of scientific work: empiricality. *Empiricality* means that the data and observations must be acquired through the senses; we get in touch with the data by experience, and we should record the data and observations with the highest possible degree of accuracy. Data gathering includes both the collecting of information by observations and measurements, and the careful testing of ideas by experimentation.

*Theories* represent the major objective of scientific reasoning: the construction of explanations of how things work in the natural world, using the empirically collected data. In the process of forming theories, several additional touchstones of scientific work are illustrated: objectivity, rationality, and generality. *Objectivity* refers to the requirement that all data and observations be considered when using the data to construct theories. We are not supposed to ignore data that do not conform to our current explanations. *Rationality* refers to the ways in which connections are made between data and between the data and the reasoning processes of theory forming. Logical thinking and mathematical relationships are two ways of putting rationality to work. *Generality* refers to the requirement that our data and theories are only interesting and useful if they can be generalized, or extended in their application to other events and objects. For example, I could spend weeks counting and measuring the lengths of the hairs on my pet cat, but such a study would have no scientific value unless it could be applied to cats in general and included some practical rationale for the study.

*Shaping principles* are those conscious and unconscious values and assumptions scientists bring to their enterprise that in very important ways shape the structure and direction of their thinking. In fact, no science can be done without them. They affect the data-gathering process, and they interact in profound ways with the theory-forming processes. They are, as the opening quote from Mayr indicates, the largely "silent assumptions that are taken so completely for granted that they are never mentioned." For example, "Occam's razor" is a principle scientists frequently use in deciding between different theories. The principle is very straightforward: Where there are several competing theories, the simplest one is chosen. The chosen theory, of course, is not necessarily correct; further work may indicate in fact that it is wrong.

Another example; *Quantifiability* has often been considered an essential quality of data—that is, if you can't measure it, it isn't scientifically valid. Yet much information from the natural world is not readily quantified. For example, how could we quantify in any meaningful way the appearance of a new species in geological time? In fact, a whole host of background beliefs and commitments influence the way scientists do their work. Empiricality, objectivity, rationality, and generality are also shaping principles. The expectation of orderly behavior and uniformity in the natural world are other examples.

Ratzsch suggests that data, theory, and shaping principles interact very strongly. They form a framework that does not permit extracting or evaluating any one component without affecting the others. For example, as a particular scientific concept broadens in its power to explain phenomena in nature, the metaphysical components (shaping principles) attached to that concept gain confidence. The converse is also true; the overthrow of a broad explanatory framework in a field of science can bring down the metaphysical assumptions attached to it. We will see a prime example of this phenomenon in the next chapter, as we examine the replacement of natural theology by natural selection. But the perspective of this triangular interaction of three components gives us a contemporary view of how science operates.

## Worldviews in the Philosophy of Science

One of the most important shaping principles is the *worldview* a scientist brings to his or her field of study. In view of the fact that science is a thoroughly human enterprise, and that worldviews determine and reflect basic values and faith commitments, it should be evident that *differences in worldviews will play an important role in controversies involving science and faith issues*. The collection of data and observations, the processes of forming hypotheses and theories, and even the range of shaping principles employed are all strongly influenced by worldviews.

As I mentioned above, there is widespread naiveté among working scientists regarding the philosophical connections of their work. I have been guilty of this in the past, and my knowledge of other scientists suggests that this is a common state of affairs. Unfortunately, we get away with it because in our publications we are only expected to deal with the data and theory components of our work. The metaphysical components of our science remain unrevealed and, very often, unexamined. As a result, instead of being informed by thorough philosophical analysis, most scientists perform their work and publish their findings informed only by their particular worldviews, which often reflect outdated or poorly conceived philosophies.

Two of the more common philosophies that can be found among practicing scientists are what we could call naive positivism and New Age subjectivism. These views are also found throughout the nonscientific community; they are now part of the spectrum of worldviews of our modern culture.

### *Naive Positivism*

Positivism is a philosophical tradition that places the strongest emphasis on empirical processes for gaining knowledge. Such knowledge begins with the objective collection of data, and leads logically through the traditional hypothetico-deductive use of experimentation to firm and reliable scientific truth. Positivists further claim that since science is the only body of knowledge that strictly follows empirical processes, the only real human knowledge is scientific knowledge. In terms of our triangular view of the structure of science, positivism strongly emphasizes the gathering of data and its use in the development of theories, and ignores the role played by shaping principles.

The strong emphasis of the positivists on data and objectivity completely overlooked the involvement of presuppositions and other nonempirical matters in the process of doing science. For this and other reasons (detailed in Ratzsch's book), positivism has all but disappeared from the philosophical scene. Many scientists and nonscientists, however, continue to hold views that are consistent with positivism; these are not philosophically reasoned views, but are of the worldview quality—prephilosophical and often unexamined. Characteristically, there is the strong belief that science holds the key to solving all of our problems; and indeed, that if some other discipline of human thought does not use scientific methods, it is suspect. As with the positivists, there is a strong belief in objectivity and a lack of respect for the role of shaping principles in the pursuit of scientific explanations. Indeed, this naive positivism elevates science to the role of a religion; when this happens, science has become *scientism*—a worldview that excludes any legitimate role for Christian faith.

### *New Age Subjectivism*

A completely different approach to science and its practice was ushered in with the publication in 1962 of Thomas Kuhn's *The Structure of Scientific Revolutions.* In Kuhn's view, science does not progress gradually, correcting itself and building ever more solid structures of knowledge. Instead, the history of science has been the history of *ruling paradigms*, broad conceptual frameworks that govern the way scientists think about and do their scientific work. Most scientists just assume the correctness of a paradigm, which is always open-ended enough to provide continuing problems for research. However, they strongly resist ideas that might challenge the correctness of the ruling paradigm. It is elevated to dogma, and strongly influences the worldview of its more committed supporters. A serious challenge to a ruling paradigm comes about only by the accumulation of clear evidence of its shortcomings, and a new paradigm emerges in a relatively short time. This is what Kuhn called a *scientific revolution.*

In the next chapter, we will employ this concept as we examine natural theology and natural selection as interpretive paradigms for biological thought in the 1700s and 1800s. Judging by the highly positive response to his work, Kuhn clearly struck a nerve. Ratzsch points out that "the Kuhnian movement has placed humans and human subjectivity (in the form of values of the community of scientists) firmly in the center of science. It has emphasized that science is a decidedly human pursuit. Science is seen as no more ruggedly and rigidly objective and logical than the humans who do it." In terms of our triangular model of science, a paradigm is clearly one of the shaping principles, strongly influencing both data gathering and theory formation.

Many of Kuhn's ideas have been used to support what has been called "New Age thinking," a phenomenon described by Loren Wilkinson in his article "New Age, New Consciousness, and the New Creation." This is a cluster of ideas centered on the view that we humans are entering a new age characterized by new modes of knowing—in particular, a new kind of science that is holistic and subjective, in which the scientist and science itself are part of a great order of being that our minds are only now beginning to be able to perceive. Western science (and Christianity) are seen as old paradigms not to be trusted; new ways of thinking are being discovered, often connected with Eastern religions that place stress on being one with nature, on the need to see ourselves as part of a larger unity.

An example of New Age thinking in science is found in the work of James Lovelock. He suggests that the earth itself is an organism capable of manipulating conditions in its atmosphere, and proceeds to develop the "Gaia hypothesis" as a new approach to environmental awareness and responsibility. Needless to say, New Age subjectivism is on the opposite end of the spectrum from positivism, and is clearly not compatible with most of "establishment science"—the mainstream science that dominates our universities and research establishments. However, it is a movement that is growing in influence, particularly on the West Coast of the United States. It is also quite clearly a new worldview, one that presents, in Wilkinson's words, "a powerful alternative to Christian faith."

### *An Evaluation*

It should be sufficiently clear that "the scientific method" is a textbook simplification that obscures the way that scientists do their work. In fact, scientists approach their work from a broad range of worldviews, and often bring to their science some very outdated philosophical assumptions. These assumptions and worldviews are most easily seen when scientists make pronouncements in defense of science or of their own work, or when they attempt to extend science into other areas of human thought. This is particularly crucial for the biological sciences, where many issues involving human life and thought are raised. It is time now to look more closely at biology as a science.

## Unique Nature of Biology

Students of biology must also study mathematics, physics, and chemistry, which speaks to a basic unity of the sciences. But there are important differences between biology and the physical sciences, differences that extend well beyond biology's focus on living things. For example: In DNA, living things have a genetic program inherited from the past. The program contains, in code, the information that generates the characteristics of the organism and (with minor errors) the means for its faithful reproduction. There is nothing comparable to this in the physical world. As a result of this inherited genetic program, living organisms have a history. A species, for example, is unique in its genetic program, the environment it has occupied, the mutations and adaptations that may have occurred in the past.

Because of these and other differences, biologists resist the assertion sometimes made by physical scientists that biology can be reduced to physics and chemistry. Living systems are far more complex than nonliving; and although it is true that none of the events or processes we study in the living world are in conflict with the described laws of physics and chemistry, the complexity of organization of living systems insures that additional characteristics emerge with each higher level of organization.

Biology is also perhaps a more disturbing science than physics and chemistry. It reaches more deeply into human life and thought, and promises to do so even more in the future. The challenges to faith, and indeed the profound influence on thinking and practice, justify the use of the term "revolutionary" in reference to biology's impact in recent times. I have chosen four revolutions for consideration in this text. The *Darwinian revolution* was highly important in providing the fundamental framework of biology and in its influence on so many areas of human thought. The *biomedical* and *genetic revolutions* demonstrated the power over the living world that comes with biological knowledge, and the consequences of that power as it challenges our ethics and social practices. The *environmental revolution* is just beginning. It is a response to the crises of population growth, pollution, resource use, and biological depletion, and confronts us with questions of environmental ethics and justice. These four revolutions have already brought fundamental changes in the way we think and live, and have immense consequences for the future. Biology is no longer the harmless conveyer of information about plants and animals it once was; it has lost its innocence and has become a catalyst for dramatic changes in the lives of all humanity.

### Theories, Concepts, Models

We have established that the task of biology is to provide natural explanations for the phenomena concerned with structure, function, and history of living systems. Assuming that the requirements of good science were met in gathering information, the critical step is to interpret the information and integrate it into the framework of existing knowledge. If we were dealing with the physical sciences, we might describe a progression from hypothesis to theory to law as a given explanation or set of ideas moves from an initial tentative status to becoming increasingly well established as valid. Laws play an important role in the physical sciences: gas laws, laws of gravity, laws of thermodynamics, and so forth. These laws faithfully describe cause-and-effect relationships, usually in mathematical terms; predictions can be made from these with a high degree of confidence. Although quantum physics informs us that even the laws of physics are statistical phenomena, we may assume that the events they describe are highly predictable, for they deal with the outcome of a large number of events. Indeed, these laws are assumed to be invariable.

However, explanations in biology do not lead to the development of statements with such high predictability that we feel comfortable with calling them laws. Instead, as Harvard biologist Ernst Mayr has pointed out in his important book *The Growth of Biological Thought*, our hypotheses and theories are organized into what we call *concepts*. The reason for this is that living systems are different in kind from nonliving systems, and far more complicated. Our explanations are always probabilistic in nature rather than invariable—for example, a swift young rabbit has an 85 percent (plus or minus 10 percent) chance of escaping from a fox once it is running. This is clearly not a law; it explains data that have been collected, and it is predictive, but only in a highly probabilistic sense.

Given this situation, according to Mayr, the history of biology is really a history of the development of new concepts—concepts such as that of the species, of homeostasis, of natural selection, of the cell—and the refining, correcting, and even rejecting of these concepts. Every branch of biology is built on key concepts, with new information feeding into the mass of knowledge on which the concepts are based. Mayr's "concepts" certainly qualify as what some philosophers of science call "theories," and others might call "models."

### *Truth in Science*

However we define the structure of biological knowledge—concepts, theories, models—it is important to consider if they are "truth," or whether our concepts or models can actually be proven true or false. If you recall, I felt rather strongly that our statement about ribbed mussels and bacteria was true, that we were in fact reporting reality when we published a paper on the research. Consider the information in your biology text;

much of it is unquestionably accepted as truth. For example, our knowledge of pathogenic (disease-causing) viruses and bacteria leads to specific immunization procedures required by law, and whole societies are protected from the diseases. Surely this information is true! Our excursion into the philosophy of science would cause us to say that there is a very good chance that it is true, but that there is no way of being absolutely sure.

We are all aware that some theories or concepts are based on limited data; we suspect that their truth is not at all well established—for example, theories dealing with the origin of life. This suggests a common-sense approach to scientific knowledge: *Our confidence in that knowledge should be proportional to the evidence supporting it.* The philosophers tell us that it is not possible either to prove or disprove any theories such that we can be sure we have them right. And it is not possible, therefore, to know whether or not our knowledge about some part of the natural world is the final word.

Where does that leave us? On the one hand, it is appropriate to accept a given explanation about the natural world as if it were true where that explanation has been reasonably well established—indeed, it would be unreasonable not to do so. On the other hand, we are wise enough to know that not all scientific explanations are equally well supported by evidence, and so we are not compelled to accept any and all explanations as valid just because they come clothed in the garments of science. This is a good occasion for us to look into the limitations of scientific inquiry.

### *Limitations of Science*

Our scientific knowledge has brought such powers through its technological applications that we are tempted to believe that there may be no limitations to what science can do. We'll see in later chapters that this impression is not true. There are four categories of limitations that we should keep in mind:

1. *Presuppositions*. The very foundational assumptions on which the practice of science is built cannot be established by science—neither scientific methodology, nor the presuppositions. For example, the orderliness of the natural world—the uniformity of nature—must simply be assumed. It is true that there are grounds for believing the presuppositions to be true, but those grounds are outside of science. This limitation suggests yet another limitation: that science cannot be the only legitimate grounds for believing something to be true.
2. *Proof*. Science cannot provide proof of its results. Scientific theories are always less than absolutely certain. They may appear to explain the data particularly well, but there is no guarantee that all possible theories (which may be infinite in number) have been considered. For example, in the early years following the discovery of the structure of DNA by James Watson and Francis Crick, there was broad agreement on what was called the "central dogma"—the concept that information flows from DNA to messenger RNA to protein synthesis, and not ever in the reverse direction. In fact, later work revealed that the central dogma was wrong—numerous cases have been found where DNA structure is determined by RNA and certain critical enzymes.
3. *Domain limitations*. Science cannot give ultimate explanations for the origin and existence of the universe (although some scientists may try). In the light of chapter 2, we see that science cannot investigate the coming into being of God's Creational Laws since God, in his creative activity, was not "bound" by the laws that he called forth. Neither can it address questions concerning the purpose of the universe or our existence. Science is also unable to speak to questions of value and morality, or of many areas of human experience such as love, honor, justice, suffering, and so forth. It is appropriate only for pro-

viding explanations of the internal affairs of the natural world.

4. *Human involvement.* Science is a thoroughly human enterprise. If we take a critical look at how it is practiced, we can see that it reflects the same kinds of problems found in all human enterprises. It may suffer from *subjectivity* (as in the choice of research and in theories we wish to examine, in data we exclude from consideration); *unethical practice* (dishonesty, for example, in reporting results); *control of the agenda* by funding agencies; *misuse of knowledge*(as in germ warfare), and so on.

   Science may also demonstrate the positive characteristics of human enterprises. There is *faith* (in the work of others, in an orderly universe); *honesty* (and it must be said that honesty and impartiality far outweigh unethical practice in science); *community* (membership in a fraternity of colleagues, sharing of methods and findings, common objectives, and so forth); *public scrutiny* of the information (this is one strong guarantor of honesty and accountability, as well as a check against sloppy work); and there is *strong commitment* to the goal of building the fund of knowledge (often in areas that are critical to the health and welfare of society).

## Christians in Science

For some reason, it never occurred to me to question whether science—and biology in particular—was a legitimate activity for a Christian. I am aware that science is regarded with suspicion in some Christian circles, and I know that biology in particular has a reputation in those circles of being far too familiar with evolution. However, I can look back and see that God used all sorts of circumstances to call me to a life of teaching and research in biology.

I was brought up in a Christian home, became a Christian at

an early age, attended an evangelical church, and went to a Christian college for two years. But the strongest influence on my career was a strong attraction to the natural world (God must have put that in me, because I can't trace it in my heredity or environment). I loved being in woodlands and wetlands; I watched birds, fished, hunted, and collected butterflies, turtles, and snakes. My feet were always wet when I came home for supper.

Yet a life in science was the farthest thought from my mind when I left high school and spent four years in the Navy. God had his plan, however; and when I entered college, it seemed only sensible to me to follow through with my basic interests and try biology. To be truthful, it has taken me many years to begin to understand what it means to see the natural world as God's creation. It has been a wonderful pilgrimage for me, taking me to strange places and into unusual experiences that have enormously enriched my life. And I can now see all sorts of reasons why Christians should go into science, and particularly into biology.

For one, doing science can be very enjoyable—the act of discovery, the opportunities to share experiences with others, the delight at seeing others use our ideas and expand on them. Another reason for going into biology is that biology is an excellent vehicle for gaining knowledge about how the creation works—what it will take to care for it so that future generations will have the same advantages and enjoyment of it that we do. A third good reason for doing science is that God has given us the responsibility of developing a culture, of learning to use the creation responsibly to form a human society that will express all of the good potential that exists in both the human mind and in the creation. One very important part of that responsibility is the abilities we gain through medicine and agriculture to ease hunger and suffering. Another good reason to do science is that we learn of God's wisdom as we uncover what he has done, and so we can give him the glory. These are just a few general reasons for doing science.

A more specific reason would be that God may have given you the gifts—and the calling—to do his work in some enterprise that involves scientific understanding. If that is true, then you had better do it! You are on a pilgrimage as I have been, and you might find some consolation in knowing that the biological path can be amazingly diverse and enjoyable as you struggle with course work and life choices.

## SUMMARY

As an example of how research leads to new information in biology, some work is described that showed the ability of ribbed mussels to feed on natural bacteria in salt marshes. This work illustrated the principle that most scientists pay little attention to the philosophical contours of their research efforts. In fact, most textbooks present an outdated view of how scientists think—the so-called "scientific method." A contemporary view suggests that three major components to the structure of science interact very strongly in carrying out scientific work: *data, theories,* and *shaping principles*. Most descriptions of scientific reasoning emphasize the interaction of data and theories, but in reality no science is done without the thorough involvement of shaping principles. These are the extrascientific values and assumptions, background beliefs, and commitments that strongly influence both data gathering and theory forming.

Worldviews represent a very important class of shaping principles; differences in worldviews are often responsible for controversy in science-faith issues. Two worldviews commonly held by scientists and nonscientists are *naive positivism* and *New Age subjectivism*. The former underestimates the role played by shaping principles; the latter downplays the power of data gathering and theory formation in understanding the natural world.

Biology is uniquely different from the physical sciences because of three factors: (1) the differences between the living and the nonliving; (2) the outcome of the theory-forming process as *concepts* rather than *laws*; and (3) the depth with which biology penetrates into human thought and life. Biology can be justifiably described as a revolutionary discipline. Four revolutions are presented; these form the subject matter of later chapters.

Although science cannot lead to absolute truth, it is reasonable to accept explanations as true when the evidence supports them sufficiently well. Limitations of science should continually be kept in mind. These are traced to presuppositions, lack of ability to prove theories, domain limitations, and human involvement.

The chapter concludes with the suggestion that science is an exciting and appropriate area for the involvement of Christians.

*Chapter 4*

# RELATING SCIENCE AND CHRISTIANITY

By every criterion laid down by Kuhn there was a paradigm of systematic natural history. Emerging from the scientific achievements of Ray, Tournefort, and Linnaeus, it involved commitments on all the levels—cosmological, epistemological, methodological, etc.—mentioned by Kuhn. Embodied in manuals and popularizations, articulated with increasing precision, communicated by precept and example, celebrated in prose and verse, it dominated the field of natural history for nearly two hundred years and helped to prepare the way for a far different, far more dynamic kind of natural history.

JOHN C. GREENE

## Some Biological History

Recognition of the important role of paradigms and revolutions is a very recent development (Thomas Kuhn's *The Structure of Scientific Revolutions* was published in 1962). Lately, philosophers and historians of science have applied Kuhn's analysis to all of the sciences, with interesting and fruitful results. Historian John Greene has examined a highly important period in the history of biology for evidence of the interaction of paradigms, the seventeenth through nineteenth centuries. He helps us gain some perspective on the sources of the problems and conflicts between biology and Christianity.

### *Paradigm of Natural Theology*

It is the testimony of Scripture that the natural world—the creation—is a source of revelation (Rom. 1:20). In accord with this, medieval thought—especially that of Thomas

Aquinas—began to erect a system of knowledge of God known as *natural theology*. Broadly speaking, the term refers to any knowledge of God apart from biblical revelation; our use here, however, will be confined to *knowledge of God as it is derived from nature*.

During the golden era of the physical sciences—the "scientific revolution" of the sixteenth and seventeenth centuries—natural theology was a highly regarded argument for Christianity that led from the observation of design and purpose in nature to the inevitable existence of God, the intelligent designer. Interestingly, at the same time, the founders of the scientific revolution were impressed with the apparent mechanical behavior of the universe as embodied in the "natural laws" they were describing. To them, God was the designer who, as a great architect or clockmaker, set things up in such a way that nature would continue to operate in accord with the fixed laws. Purpose meant ultimate purpose, an answer to the question "why" rather than "how." It was the domain of science to speak to the question of *how* nature operates.

However, the work of these founding fathers of physics and chemistry had little immediate influence on biology, which during the seventeenth and eighteenth centuries was concerned with two major areas: medicine and natural history (natural history refers to an approach to studying the natural world that emphasizes the discovery, description, and classification of the things of nature). The natural historians—men like John Ray (*The Wisdom of God Manifested in the Works of the Creation*, 1701) and Carl Linnaeus (*The Oeconomy of Nature*, 1791 and *A General System of Nature*, 1806)—were clearly committed to natural theology. A quote from Linnaeus's *Oeconomy* gives the flavor of this commitment:

By the Oeconomy of Nature we understand the all-wise disposition of the Creator in relation to natural things, by which they are fitted to produce general ends, and reciprocal uses. All things contained in the compass of the universe declare, as it were, with one accord the infinite wisdom of the Creator. For whatever strikes our senses, whatever is the object of our thoughts, are so contrived, that they concur to make manifest the divine glory, i.e., the ultimate end which God proposed in all his works. [After declaring this truth, Linnaeus applies it to the study of what we today would call ecology.] . . . In order . . . to perpetuate the established course of nature in a continued series, the divine wisdom has thought fit, that all living creatures should constantly be employed in producing individuals; that all natural things should contribute and lend a helping hand to preserve every species; and lastly, that the death and destruction of one thing should always be subservient to the restitution of another. It seems to me that a greater subject than this cannot be found . . .

As John Greene comments in the opening quote to this chapter, this is clearly a paradigm, and it dominated the field of natural history for almost two hundred years. According to this paradigm, the organisms inhabiting the earth were part of the original creation, and their adaptations were seen as the result of providential design. This latter concept is known as *teleology*—the concept that, as Aristotle put it, "there is a purpose in what is and what happens in nature." Confronted with, say, the remarkable way that some insects resemble the twigs or leaves on which they rest, the teleological explanation for the adaptation would be that God designed it that way when he created the insects in order to provide for their protection.

In the paradigm of natural theology, design and purpose became the explanation for observations in natural history. Every evidence of adaptation, of the amazing correspondence between form and function in living things, spoke of design. And the existence of such wonderful and obvious design argued strongly for the existence of a Designer. The function of natural history was thus to explore and describe and classify the things in nature in order to admire and praise the skill of God, the intelligent Designer. There was no search for mechanisms that could explain how adaptations might have come about, and in this sense the natural historians were unlike their physical science counterparts. It was a static paradigm in which the natural historians were saying *why*, and thought they were describing *how*. Design and an original creation by God answered all of the questions about origins and adaptations. But the dominance of natural theology as a paradigm for doing natural history was to end, and the effects of its demise were far-reaching.

### *Seeds of Conflict*

The new paradigm did not arise from within natural theology as a result of a growing sense of dissatisfaction with the old paradigm. Indeed, it was not immediately recognized as a rival when it first appeared in tentative and incomplete form. The Comte de Buffon's *Histoire Naturelle* was published at around the same time as the works of Linnaeus (1749), but Buffon's approach was radically different from that of Linnaeus. Where Linnaeus sought to classify the plants and animals according to a pattern that he believed existed in the mind of the Creator, Buffon felt that taxonomy was an artificial imposition on what he saw as a confusing array of living forms. To him, the primary purpose of natural history was to explain observations of living things on the basis of the actions of natural laws and events. Buffon was strongly influenced by the revolution that had occurred in physics and cosmology—and, it must be said, seemed little inclined to see design as an explanation for what he discovered. Instead, he proposed a new kind of natural history—one that sought natural causes, was dynamic, nonteleological, and comprehensive. His work was the first of many to look beyond design as a way of explaining adaptations in the living world.

Georges Cuvier, in the early nineteenth century, was concerned about the widescale extinction of species that geological work had uncovered. Natural theology had assumed "the works created by God at first, and by him conserved to this day in the same state and condition in which they were first made." Cuvier postulated successive creations by God over time to account for extinctions, and so stayed within the paradigm of natural theology.

During the same time period, Jean Baptiste Lamarck undertook a complete revision of biology and geology from a perspective that incorporated the following elements: *geological uniformitarianism* (present processes may be extended into the past to explain the past); a *vast time scale; descent of organisms with modification; and progressive development* of organic forms from simple organisms up to human beings—in short, most of what Darwin would a short time later describe in detail in *The Origin of Species.* Lamarck was a thorough evolutionist, but he did not have a theory of how species could originate, and his views on how the environment affected heredity were erroneous.

Like some would do after Darwin's work, Lamarck asserted that God endowed the natural world with its evolutionary abilities: "Nature, in successively producing all species of animals, beginning with the most imperfect or the simplest, and ending her work with the most perfect, has caused their organization gradually to become more complex." This trend, he explained, comes "from powers conferred by the supreme author of all things. Could not his infinite power create an order of things which gave existence successively to all that we can see as well as to all that exists but that we do not see?" Even though his work seriously challenged the static view of nature held by natural theology, Lamarck failed to make a convincing case; it remained for Charles Darwin to do that.

### The Demise of Natural Theology

Like Lamarck and Buffon, Darwin offered a new approach to natural history that differed radically from the ruling paradigm of natural theology. Instead of an original, extensive creation of fixed species, or of successive creations, Darwin presented evolution—descent with modification. Instead of seeing adaptations as the result of design by a wise Creator, Darwin saw adaptations by natural selection—the creative force behind both the adaptations and the changes in species over time. Darwin's work delivered the death blow to the ruling paradigm by offering a plausible explanation for natural history from within the natural world. The new paradigm became a highly preferable substitute for the old for scientists who, by the mid-nineteenth century, were ready to accept a view that could explain the phenomena of the biological world on the basis of the operation of natural forces and laws. This paradigm—the *Darwinian revolution*—is dealt with more thoroughly in chapter 7. Our focus here is on its impact on natural theology.

Darwin so thoroughly presented the case for evolution in *The Origin of Species* (published in 1859) that most of his true audience—the active scientific community in Britain—quickly accepted this new explanation. The symbolism of conversion was often used in reference to this process. Darwin understood well what he was doing to natural theology. Writing in his autobiography, he said,

> The old argument of design in nature . . . which formerly seemed to me so conclusive, fails, now that the law of natural selection has been discovered. There seems to be no more design in the variability of organic beings and in the action of natural selection, than in the course which the wind blows. Everything in nature is the result of fixed laws.

The new paradigm in natural history—Darwinian evolution—thrust a dividing wedge between science and religious belief that continues to this day. The authors of the scientific revolution in physics and cosmology—most of them strong believers—had unwittingly paved the way for this development by presenting the view of an autonomous nature bound by fixed laws and functioning without any direct reference to God. They saw God as the creator of the universe and the physical laws, and also designer of the living world, all of which took place, supernaturally, at the original creation. Darwin's work basically removed creation and design from the supernatural realm and placed it into the realm of autonomous nature, as law-bound as the stars in their courses or the mechanics of gravitational force. In doing so, he joined the emerging forces of higher criticism in challenging the historicity of early Genesis.

Here, then, was a major paradigm shift in the Kuhnian sense. The theoretical components of Darwinian evolution provided a far more satisfying explanation of the biological and geological data than had the natural theology paradigm. And because the prevailing theoretical explanations were so intimately tied to theistic assumptions, the Darwinian revolution had the appearance of calling into question some of the most fundamental Christian beliefs of the day. In terms of the three components of science outlined in Chapter 3, here was a clear interaction between *theory* and *shaping principles* that led ultimately to the overthrow of a major system of interpretation. By linking Christian themes so strongly to an interpretation of data from the natural world, the proponents of natural theology had unintentionally put at risk the broadly accepted belief (in the Western world) in a creator-God.

It was not long before the theological community began to respond to Darwinism. The earliest responses were basically expressions of outrage—science was making unreasonable claims that were clearly incompatible with biblical teaching. Highlighted by debates in Victorian British society, the issue at first centered on the reliability of the biblical account of creation. The first serious treatment of Darwinism from orthodox Christianity came in 1874 from Princeton theologian Charles Hodge. Hodge studied Darwinism thoroughly and grasped its implications for natural theology. At the end of his book, *What is Darwinism,* Hodge asserted that "The conclusion of the whole matter is that the denial of design in nature is virtually the denial of God." For Hodge, Darwinism was not an option for Christian belief; either Darwin was wrong, or else God did not exist. Natural theology had clearly made its inroads into Christian apologetics.

### A Critique

Whose side shall we endorse here? I would suggest that natural theology and the argument from design, as a paradigm for natural history, deserved to be put to rest. It was clothed in a static view of the natural world; although it dealt with the concept of purpose in the created order, design had become a "scientific" explanation for adaptation. Just as God had set in motion the stars and planets, and had established natural laws for the operation of matter and energy in the universe, he had also built into (designed) living organisms when he created them the adaptations they needed to function in different natural environments. This view has much in common with *deism*—the belief that God initiated all things and then stepped back and allowed the natural world to function automatically.

As we mentioned in chapter 2, the biblical view does not attempt to establish "natural laws," nor does it attempt to describe how God might go about designing his creatures. Instead, we find creation by God's *Word*—the same Word that, as Psalm 147 testifies, flashes to earth to spread snow and frost and ice. It symbolizes God's *authority* over the natural order. We also find creation by *Wisdom*, symbolizing the carrying out of God's *purposes* in the development of the created order over time. The natural theology paradigm for natural history was heavily based on the assumption of an original supernaturalistic creation that included the origin of adaptations. This static view of nature was obviously vulnerable to an alternative explanation that was rooted in the natural world and presented in a convincing fashion, as was Darwinian evolution.

As a paradigm for study, natural theology had functioned well enough to earn grudging praise from later evolutionists; it had laid a solid foundation for evolutionary biology. The natural historians had done well in describing the amazing ways in which God had designed the relationships of organisms to each other and their environment—their adaptations. They were fascinated with the diversity of plants and animals to be found, and were instrumental in establishing the fields of taxonomy and ecology. Indeed, most of the literature from this era could be incorporated into evolutionary thought simply by replacing "the hand of the Creator" with "natural selection" in the generation of explanations for adaptation. This simple substitution took the living world from the level of teleological explanation to the level of a mechanism that would give a reasonable account of how adaptations came about. In doing so, however, it gave the appearance of taking God's work away from him and at the same time casting serious doubt on the accuracy of the Bible.

Because of this and other conflicts, historians of science have commonly proposed that biology (in the form of Darwinian evolution) and Christianity were engaged in virtual warfare during the latter half of the nineteenth century. Things were not so simple, however. The response from orthodox Christianity, first clearly articulated by Hodge and later picked up by more fundamentalist theologians in the twentieth century, was clearly antagonistic. However, there were trends toward a reconciliation of evolution and theology in several theological traditions, including Roman Catholic, Reformed, and liberal Protestant. Further, there were at that time numerous scientists who were Christian believers (such as the prominent Harvard botanist Asa Gray), who sought to bring their faith to bear on science in a spirit of compromise.

In recent years, historians of science have agreed that the

metaphor of warfare between science and Christianity is inappropriate. It would be more accurate to speak of "encounter" rather than war, and encounter in numerous and complex ways and for a variety of reasons. The net result of this encounter has been a growing recognition that science and religion are not necessarily incompatible; that when the limitations of science are recognized and when the claims of faith are made clear, these two areas of human endeavor have much to say to each other. The nineteenth-century upheavals served to remove religion from scientific theory forming, where it did not belong. In the process, Christian thinking about God's relationship to his creation has been strongly affected and, in the view of many, greatly clarified. But our historical discourse would not be complete if we ignored the worldview implications of the encounter between Christianity and science.

## Competing Worldviews

The real and enduring conflict is between rival worldviews. The paradigm of natural theology accompanied a worldview that was at its best theistic, and more commonly, deistic. Most people acquainted with Christian doctrine in the mid-1800s shared the common beliefs of Christians about the existence of God and other major elements of Christian thought—whether or not they had a personal faith. Those made uncomfortable by notions of a God who might stand in judgment and a faith that took biblical teaching seriously could embrace deism, whose God was a remote clockmaker who wound everything up in the beginning and then remained in the background.

By challenging the static view of natural theology, Darwin was also challenging the worldview in which it was couched—and for a time it was his own worldview. Shortly after returning from his voyage on the *Beagle*, Darwin began to examine the grounds of Christian faith. As he did, he found his own beliefs changing. While he was forming his concept of evolution by natural selection, Darwin saw the argument from design disappear. He did not enjoy the controversy that followed publication of *The Origin of Species*, and he expressed toward the end of his life an agnostic view: "Let each man hope and believe what he can." So along with natural theology, in the minds of many of Darwin's day, also perished traditional belief in the authenticity of the Bible and therefore the basis of Christian faith. Much of the opposition to Darwinian evolution that persists to this day can be traced to this nineteenth-century attack on the Christian worldview.

One scientist very quickly grasped the worldview implications of Darwinism and exploited them with enthusiasm: biologist Thomas H. Huxley. Huxley believed that the fall of natural theology meant the removal of God as creator and designer, and therefore the loss of nature as evidence of God. Already convinced of the fallacy of biblical Christianity, following publication of *The Origin of Species* Huxley quickly turned his sharp wit and debating skills to the task of completing the overthrow of Victorian Christianity. Huxley believed that science represented a new, revolutionary vision, one that was not based on revelation but on cold reason. He coined the word *agnostic* to express his new faith, and described it as follows:

> Agnosticism, in fact, is not a creed, but a method, the essence of which lies in the rigorous application of a single principle. . . . Positively, the principle may be expressed: In matters of the intellect, follow your reason as far as it will take you, without regard to any other consideration. And negatively: In matters of the intellect, do not pretend that conclusions are certain which are not demonstrated and demonstrable. That I take to be the agnostic faith, which if a man keep whole and undefiled, he shall not be ashamed to look the universe in the face, whatever the future may have in store for him.

Huxley proposed that science could take the place of revelation as the source of truth and the foundation for society's future. He spoke in Christian symbolism of conversion (to agnosticism and Darwinism), of the "church scientific," of scientific miracles, and so forth. Writing to Charles Kingsley, a clergyman, Huxley asserted,

> Understand that all the younger men of science whom I know intimately are essentially of my way of thinking. . . . Understand that this new school of prophets is the only one that can work miracles, that it is right and you will comprehend that it is of no use to try to barricade us with shovel hats and aprons, or to talk about our doctrines being "shocking."

### *Scientific Naturalism*

Huxley was obviously endorsing a new worldview that rejected biblical Christianity on scientific grounds. Darwinian evolution, as we will see in chapter 7, is properly examined on the basis of its ability to explain the data collected from a number of scientific disciplines: geology, ecology, taxonomy, anatomy, molecular biology, and so forth. But it is also quite readily drawn out of its context as a scientific explanation and used in support of a worldview—*naturalism*—as Huxley was doing.

This view holds that naturalistic explanations provide all we need and indeed are able to know; ultimate reality is matter and the space-time arena in which it operates. This is Nature with a capital N, and it employs evolution as a substitute for divine action. Evolution creates, designs, and exists without reference to anything external; it is self-governing, in a self-governing system of matter and energy that is the cosmos. In this way, scientific naturalism clearly makes judgments on value and purpose in the natural world (namely, that there are none), and declares the autonomy of the existing world of Nature.

E. O. Wilson is a modern-day Huxley who discusses the relationship between science and religion in his book *On*

*Human Nature*. In Wilson's view, the new science of *sociobiology* is able to explain the phenomenon of religion (a "mythology"), which is the chief competitor of science: "The scientist's devotion to parsimony in explanation excludes the divine spirit and other extraneous agents. Most importantly, we have come to the crucial stage in the history of biology when religion itself is subject to the explanations of the natural sciences . . ."

In my view, the task of Wilson and others who elevate science to worldview status is made easier by some of the attempts made in the name of Christianity to discredit evolution. Such attempts make it appear that the only worldview alternative to scientific naturalism is what has been called *special creationism*. Once Genesis is thought to be limited to the literalistic interpretations of this worldview, it only remains for the defenders of evolutionism to show the shortcomings of scientific creationism. They can ignore other theistic views and declare an easy victory.

It is apparent that history is repeating itself, that the triumph of Darwinian evolution over natural theology has its modern counterpart in the contest between scientistic naturalism and scientific creationism. Once again, the conflict is really a conflict of worldviews, and therefore has dimensions that reach into matters of belief and ultimate reality. In a real sense, the minds of a new generation are a battleground for allegiance; will it be science, or God? To address this matter further, let's examine more directly the relationship between science and biblical truth, and begin with a look at what it means to consider the Bible as a source of truth.

## Scripture and Science

### *Interpretation of Scripture*

It is my belief that the Bible—both Old and New Testaments—is the written Word of God (in the original texts), and is both reliable and authoritative as a source of information about God and his works. The Bible can be said to have both human and divine authorship; in other words, God used human means and language to convey his revelation to us. "All Scripture is God-breathed and is useful for teaching, rebuking, correcting and training in righteousness, so that the man of God may be thoroughly equipped for every good work" (2 Tim. 3:16,17). The Bible therefore is much more than a sourcebook for theological studies; as the 2 Timothy passage indicates, there is a central purpose to Scripture: *to convey an understanding of God that leads to salvation through faith in Christ, and to equip the Christian for a life of good works.* This is the doctrine of the inspiration of Scripture, and it has always been considered a crucial doctrine for orthodox Christianity. However, it is one thing to say that the Bible is inspired, and another to then turn to Scripture and interpret what it is saying.

The Bible was written by numerous authors over a span of 1,500 years, ending some 2,000 years ago. The doctrine of inspiration holds that, in the process of writing, the authors were conveying God's message, his revelation; but it is clear that the message was imbedded in the thought patterns and vocabulary of the authors. Where, for example, biblical writers referred to historical events, such as the events in the life of Christ, each writer showed significant individuality in the details and the sequence of events. Underlying the record, however, is the historical reality they were reporting. And the framework of their writing was the culture and worldview of their times. Therefore the process of interpreting Scripture must recognize the framework and the individuality of the authors.

Interpretation of Scripture is, in one sense, the privilege—and responsibility—of all who come to the Bible with a desire to know what God has to say. But, in another sense, interpretation can become a highly demanding, scholarly activity known as *exegesis*. Here the scholar approaches the Bible with certain principles of interpretation (hermeneutics), such as a study of the original language to catch subtle differences in word meanings, a careful recognition of the context of a passage, and a thorough appreciation of the history and culture of the people to whom and by whom the book was written. This process operates out of a very definite theological paradigm, and there are conflicting paradigms depending on other basic theological presuppositions and beliefs.

In short, interpretation of the Bible is both a scholarly and a lay activity, to which the reader brings presuppositions and a methodology in the search for the meaning of a part of Scripture. And it is a twofold task: first, to discover what the message meant to its original audience; and then to determine what it has to say to us in our culture. Because of all of these considerations, it is not surprising to find well-meaning Christians disagreeing about the interpretation of a passage of Scripture. Yet it is true that an understanding of the meaning of Scripture is essential to settling disputes about the relationships between science and the Christian faith.

### *Interaction of Scripture and Science*

How shall we deal with issues that bring the Bible and science into contact? This is one of the most important questions facing us as we look at biology through the eyes of faith. To stimulate our thinking, let us pose two propositions:

1. The Bible contains information that leads to generalizations applicable to science.
2. The observations and experiments from the natural world lead to generalizations with theological content.

Those early Christian biologists who brought us the paradigm of natural theology were convinced of both of these propositions. Their reading of early Genesis told them about how natural things began and how they happened to be so well adapted, and their studies in the natural world brought them such strong evidence of design that it was only logical to point to God as the origin of that design.

However, the subsequent fate of the natural theology paradigm and the broad impact its demise had on human thought makes it clear that we need to examine carefully this possibility of dialogue between the Bible and science. What, then, of our first proposition, that the Bible contains data that lead to generalizations applicable to science? We established in chapter 2 that the Bible has much to say about the natural world as God's creation, and speaks of his continuing action in upholding all things as the Creational Lawgiver. Scripture also assigns value to the things of creation, and assures us that this creation shows God's glory to us.

My understanding of the structure of science indicates that such biblical principles may play a legitimate role as framing principles, but that they are not data, nor are they theories in the scientific sense. They may influence the way we do our science, but that is a very different matter from postulating that the Bible contains relevant scientific data—information that not only has status as a source of information for scientific theorizing, but because of the Bible's authority as God's Word, possibly even takes precedence over data from the natural world. In the next chapter we will examine Genesis 1, which many Christians believe does indeed contain scientific information.

Here are three options for the impact of the Bible on science:

1. *Scripture is scientifically accurate and in fact contains the best source of information about some aspects of the natural world.* Advocates of this view hold a very literal interpretation of the Bible and, it must be said, construct a science that is at odds with contemporary science in many areas (in particular, the age of the earth, the mode

of development of life forms, the extent of the flood, and so forth).

2. *Scripture is scientifically accurate, and should harmonize with the findings of science.* Those who hold this view have a high regard for the results of science, and seek for a harmonization of science with scientific information found in Scripture.
3. *Scripture does not contain scientific data.* In this view, the intent of Scripture is to convey knowledge about God and his works, but this is not to be seen as scientific data. Passages that might appear to have scientific implications must be interpreted in the light of the worldview and life view of the writers. We need to direct our thinking to the theological core message of such passages. Such a message may indeed have relevance to science, but the relevance relates to such areas as meaning, purpose, and status, to which science does not directly speak.

These three options form the basis of some well-established models for relating science and biblical truth, which we shall outline in the next chapter.

### *Does Science Inform Theology?*

In considering the relationship between science and Scripture, we have looked at the possibility that the Bible contains data relevant to science, the first of our two propositions. But what of the second proposition: that the observations and experiments from the natural world could lead to generalizations with theological content?

As a first response, recall chapter 2, where we found that our studies of the natural order—the creation—should inform us about the wisdom, power, and glory of God; and that its highest purpose would be to put us in touch with the Creator-Redeemer, Jesus Christ. If we ask the right questions—scientists or not—questions of value and purpose and governance in the natural world, Scripture tells us that those questions should lead us to God. These are not questions that science can answer; but our acquaintance with the natural world, whether it is simply appreciation or goes more deeply into explanations of how things work, should lead us to ask them. In this sense, there is theological content in the findings of science. This is natural theology as it can presently be understood.

In our look at the problems of interpreting Scripture, we noted that the context of a passage of Scripture is highly important in determining its meaning. Chemist Robert Fischer suggests that because God has revealed himself both in Scripture and in the created world, the realm of nature can legitimately be regarded as part of the context for interpreting Scripture. Both are valid sources of information, and since both are God's revelation, there should be no conflict between them. This is an important perspective; it provides us with a working faith and a goal—the goal of reconciling apparent discrepancies between these two sources of information. This perspective means that where we think Scripture is informing us about the natural world, that information should not conflict with the information provided by a true knowledge of the world itself.

However, when we consider the methods of theology and interpretation of Scripture, on the one hand, and the methods of science on the other, we understand that both of these fields have their limitations. Both are human enterprises. For this reason, we should be willing to examine our presuppositions and our paradigms in both fields, and hold our views and conclusions somewhat lightly. We should not be surprised when problems arise.

## SUMMARY

Science and Christian belief were strongly related in the early days of the scientific revolution. Natural theology was for scientists of the seventeenth through nineteenth centuries an important argument for the existence of God. The physical scientists saw design and purpose in the origin of the universe and the laws that governed it. The founders of natural history extended the argument from design to explain the adaptations of living things, and explored and described and classified organisms in order to point out the skill of God, the great designer. Although their work provided a good foundation for classical biology, their paradigm did not encourage scientists to search for mechanisms that might explain origins and adaptation.

Darwinian evolution provided a new paradigm for natural history that offered an explanation for origins and adaptation from within the natural world. This paradigm was broadly accepted in the scientific community, and its acceptance had serious impact on religious belief by casting doubt on the biblical account of creation and belief in God's work in nature. The static paradigm of natural theology failed because it confused purpose with mechanism. Evolution generated a reaction in the Christian community that can best be characterized as a continuous encounter, one that has in many ways clarified thinking about God's relationship to his creation.

At the level of worldviews, Darwin clearly ignited a conflict. T. H. Huxley believed that evolution was a triumph of reason over blind faith, and promoted agnosticism as superior to Christian faith. This use of evolution to promote a naturalistic worldview is continued in the modern-day work of biologists like E. O. Wilson. The promotion of an alternative theistic worldview is made difficult because of attempts by some Christians to discredit evolution.

A contemporary look at science and Christian belief begins with an understanding of the origin and purposes of the Bible. Some ways of interpreting Scripture assume that the Bible is a source of scientific information, and generate serious disputes with science. Other approaches see biblical truths as being related to science as important framing principles, but not as part of the data-gathering or theory-forming process.

Scientific work can impact Christian thought by heightening our awareness of God and his works, and also by helping us to clarify our interpretation of Scripture. It is an important conclusion of faith that both science and Scripture are sources of knowledge of God's works and that, when properly understood, should not lead to conflicts.

*Chapter 5*

# PERSPECTIVES ON GENESIS 1

Nothing is here treated of but the visible form of the world. He who would learn astronomy and other recondite arts, let him go elsewhere. . . . It must be remembered that Moses does not speak with philosophical acuteness on occulte mysteries, but states those things which are everywhere observed, even by the uncultivated, and which are in common use.

JOHN CALVIN

## Interpretation of Genesis 1

### *Questions*

How shall we understand Genesis 1? Is it science, history, theology, poetry, saga, myth, or what? This question is one that any biologist who is a Christian must answer sooner or later. Inevitably, answering this question leads to yet another question: Which is true: creation or evolution? Again, Christians in biology are directly in the line of fire, and in this case it is a crossfire. We are challenged from one direction by fellow Christians, who often use this issue as a test of orthodoxy, and from the other direction by non-Christian friends and associates: How can you be a scientist and also believe in the Bible?

Many Christian biologists caught in this crossfire are becoming weary of the battle. Yet there are good reasons why we need to address origins issues head-on in this book. Students are often confused about these issues. They want to know how to respond appropriately to the conflicting answers from their church and from their schools. And we must understand that these questions are often symptomatic of deeper questions: Is my faith real? Is the Bible true?

Another good reason for getting involved with these origins

issues is that we would like to be able to put them in their proper perspective so that we can get on with other very important problems. Cal DeWitt, director of Au Sable Institute of Environmental Studies, is fond of telling the following little story: In the midst of a large city is a magnificent building, laid out in breathtakingly beautiful lines. But as we approach the building, it becomes apparent that something is terribly wrong. There are people inside and outside the building in the act of vandalizing it. Windows are being broken; a fire is burning on one of the top floors; graffiti are being scrawled on the walls; people are running off with the furnishings. We get closer, shocked at what is happening, and then we come upon another group of people standing across the street from the building. They look quite respectable, but they are engaged in a heated debate. As we approach them it becomes apparent that they are arguing furiously about how the building was built. One group is claiming that the builder put the building up practically overnight; another group maintains that it took many years to build it. Some feel that the architect drew up his plans and then went away; others were sure that he supervised every detail of the building. We interrupt the argument and ask them why they are standing there debating about how the building was built while the whole structure is in the process of being dismantled. And they really don't have a good answer.

### *Approaches*

Our study of Genesis 1 will point up some differences between Christians in approaches, interpretations, and conclusions. We must recognize that these are differences between *believers*—people who are sincerely trying to be obedient to biblical teaching. However strongly we support a given view, we must not be tempted to judge opposing viewpoints as being non-Christian. These matters have been debated for many decades, and still the differences persist. So we must accept as a starting premise that the issues surrounding

Genesis 1 are sufficiently cloudy that no one view can be considered "the Christian view." Indeed, there are major doctrinal issues at stake; but several ways of interpreting Genesis 1 preserve these doctrines.

Our approach will be to study this passage as God's word and as human words. Genesis 1 is no less inspired than, nor is it subordinate to any other part of the Bible. Our primary objective is to interpret the message of this important passage of Scripture both for its original audience and for today. At the end of the last chapter, we suggested along with Fischer that since God's revelation is both in Scripture and in the created world, that world can be considered as part of the context for interpreting Scripture. British research chemist E. C. Lucas asks an important question: "Is it ever right to take scientific issues into account when seeking to understand Genesis 1–3 . . .?" Lucas argues that since the sciences of language and history are used as tools in interpreting Scripture, we should be able to put the natural sciences to use as well. He affirms with Calvin that all truth is God's truth, and so we should "accept gratefully, as God-given, all truth, from whatever source, that might aid us in understanding Scripture." A potential problem crops up, however: If we allow scientific findings to influence our interpretation, that would appear to be placing contemporary human knowledge in authority over God's Word.

Perhaps the best way out of this quandry is to acknowledge that both our science and the process of interpretation of Scripture are fallible human enterprises. They both lay hold of truth, some of which is better understood, and some less so. We should, therefore, recognize the limitations of both of these enterprises as we look at the various ways of relating science and early Genesis. In this chapter, we will examine the major rival interpretations that have support from different Christian groups, consider the advantages and problems posed by each, and determine what each interpretation presents as the message God intended to communicate in the passage. We will then present several broad models for relating science and Scripture and use our work with early Genesis to evaluate them.

## Genesis 1:1 to 2:3

I would suggest that you read this part of Genesis before going further, and keep it close at hand for reference. What kind of literature is it? Prose or poetry? There is clearly a rhythm to it, and a number of repeated words and phrases, yet it lacks the rhythmicity and parallelism of later Hebrew poetry (such as the Psalms and Proverbs). French theologian Henri Blocher, whose thinking we will follow closely as we proceed, explains:

> The word "hymn" comes to many writers. Whether it is a strophic hymn in prose or a hymn which is a unique blend of prose and poetry, Paul Beauchamp, the most sensitive of analysts, wisely concludes: "By the importance of repetition and of its corollary, silence, our text is indeed close to poetry, but its movement towards a solution places it in the order of prose." Following his lead we can say that the genre is composite.

Blocher points to a careful construction by the writer in which symbolic numbers—ten, three, and seven—are heavily employed. For example, "God said" appears ten times, as does the verb "to make" and the formula "according to its kind." There are three benedictions, and other groups of three in the structure. The formula "and it was so" appears seven times, as does God's approval "and God saw that it was good," independent of the seven days. Blocher summarizes: "Beyond any doubt, here we have no ordinary history, such as might be written in response to a simple request to be told what happened. Here we have the work of a Master whose thought is profound and expansive." Blocher goes on to suggest that

God inspired the writer (assumed to be Moses) in much the same way as with authors of Wisdom literature, "not by short-circuiting him but, on the contrary, by directing and bearing along his meditation. . . . the Wisdom theologian who draws up his thoughts gives us 'knowledge in a most concentrated form,' matured in meditation on the finished works of God."

### *First Impressions*

All of the interpretations of early Genesis agree that the most important message of the text is to establish God as creator, in order that the hearer or reader might respond to God's loving care in faith and love for him. The people of Israel could know their covenant God as the creator of "the heavens and the earth in all their vast array," as Genesis 2:1 puts it. God called his creation into being. The universe is not eternal—it had a beginning. And it is obvious that the framework of this creative activity is the seven-day week, where each day is carefully identified and numbered. The correspondence of the "creation week" with the institution of the sabbath is clearly intended by Moses, as Exodus 20:11 tells us.

Leaving for the moment the details of the account, it might be good to think about what is *not* said here. Nothing is said about the means or mechanisms used by God to create. I think it is fair to conclude that it was not the purpose of the author of this passage to attempt to describe God's creative techniques. Although some of the terminology seems to point to mechanism ("Let the land produce vegetation . . . Let the land produce living creatures . . ."), there is each time also the statement "So God created . . ." that links the creative activity directly to God.

It is also instructive to consider the author's language in this passage. It is the language of everyday experience, written from the author's perspective (the worldview of the Hebrew culture). It is nontechnical, nontheoretical language—it describes the way things appear to the senses of the observer. The waters are "gathered" together by God; God "sets [the lights] in the expanse of the sky"; animals and plants reproduce "after their kind" (this explains reproduction in ordinary language). Blocher explains:

> In the act of inspiration God did not turn his spokesmen into robots; his Word became their word, under their signature and their responsibility. Thus we have no right to go over their heads in order to set forth a "divine" meaning which they would never possibly have imagined—even if those men did not grasp the whole import of what they attested (I Pet. 1:10 ff.) . . . In the case of the opening chapters of Genesis, it is not plausible that the human author knew what we are taught by astronomers, geologists and other scientists.

Consider also the historical context. It is likely that Moses wrote the Genesis narratives during the wilderness wanderings of the Israelites. At this point in their history, the Israelites were a largely pagan people, lacking any written Scripture and heavily influenced by the pagan Egyptian culture they had been part of for four hundred years. Their understanding of the God of Abraham was sketchy at best, and yet they were willing to agree to a covenant with God, to become "a kingdom of priests and a holy nation." This would require that they know much more about the God who led them out of bondage, and the five books of Moses—the Torah—were God's inspired foundation for the firm establishment of the Hebrew religion and culture.

One essential component of that foundation would be an understanding of origins—and a stark contrast with the teachings of the pagan religions around them with respect to origins was crucial. Genesis 1, then, was written to these ancient Israelites to teach them that their convenant God was the God who created all things. They were to remember that creation and the Creator as they learned to keep the Sabbath; and the framework of the seven-day week with the day of rest became both the framework of their daily lives and the framework in which they could picture their creator-God. It is hard to imagine any reason why the Israelites might need to have a sequence of God's creative activity, or why they might be concerned about how he did it. However, as we will see, many Christians believe that this was part of the message of early Genesis.

### *Major Interpretations*

A great deal of imaginative work has gone into the construction of theories interpreting the six days of creation; Dallas Cain lists fourteen different interpretations offered by Christians over the last 200 years! We will look at four interpretations, all of which take the passage seriously (none consider this part of the Bible to be mythology): the *reconstruction theory*, the *day-age theory*, the *literalist theory*, and the *framework theory*.

#### *Reconstruction Theory*

According to the reconstruction theory, God created an original, perfect creation (Gen. 1:1), which could have occurred in the ancient past. Verse 2 is interpreted as stating that "the earth became without form and void," implying some great catastrophe occurred that wiped out most of the structure and life on earth. (It is often suggested that the catastrophe was the consequence of evil entering the world at the fall of Satan.) Then the six days of creation describe God's "reconstruction" of structure and living things. The interpretation here may be a literal six-day creation, or perhaps days as longer periods of time marked off by a beginning or an ending. (This theory became widely publicized by its inclusion in the *Scofield Reference Bible* in the early 1900s). The advantage of this view is that it allows an apparent accomodation to the fossil record and even to evolution. Scofield's comment:

It is by no means necessary to suppose that the life-germ of seeds perished in the catastrophic judgment which overthrew the primitive order. With the restoration of dry land and light the earth would "bring forth" as described. It was animal life which perished, the traces of which remain as fossils. Relegate fossils to the primitive creation, and no conflict of science with the Genesis cosmogony remains.

The problems with this view are serious. It takes unwarranted liberties with the Hebrew text in postulating the catastrophe of Genesis 1:2, and in assuming that elsewhere in the creation account God is "remaking" or "recreating." It also presumes a connection between Satan's fall and cosmology that is nowhere supported in Scripture. In other words, the theory exploits the silence of Scripture and erects a non-Scriptural framework for God's activities. Furthermore, although the theory was popularized as an attempt to reconcile Genesis with contemporary science, there is no evidence from the fossil record of a single catastrophic destruction and then the redevelopment of life forms. As we will see in the next two theories, attempts to blend the Genesis account with science or to postulate a literal six-day creation have their own difficulties with scientific findings.

### *Day-Age Theory*

The day-age theory holds that the "days" of Genesis 1 are to be taken figuratively and are actually long periods of time. There are numerous variations on this theme: The days may be distinct ages in series; the ages may be overlapping; the days might be literal days, where each day is followed by an age that may or may not overlap with others. These views have in common the desire to harmonize Genesis 1 with geological and biological information, although those who propose this theory are usually not comfortable with evolution. Their solution to the development of life over time is often *progressive creation*, where God creates basic "kinds" of plants and animals at different eras and on different creation "days," and the "kinds" then develop into a variety of related forms over time by microevolution.

John Wiester has presented the most recent version of this theory in *The Genesis Connection*. After a detailed presentation of current scientific views of origins in the light of the Genesis account, Wiester draws the following conclusions: (1) The order of events in Genesis 1 and the scientific record are in substantially the same sequence; (2) all major explosive adaptive radiations correlate with Scripture's creation commands (this refers to a current and debated view in evolutionary theory called "punctuated equilibrium," which we will look at in Chapter 7); (3) each creation command correlates with a scientific puzzle or gap (for example, Genesis 1:9 refers to the separation of dry land from water, and science does not have a good answer to how this might have occurred). Wiester's conclusion:"Scientists are now furnishing substantial evidence that the biblical account of Genesis One is an accurate historical account of creation. At the very least, one is drawn to the conclusion that there is no conflict between the evidence provided by modern science and Genesis One." And later on: "How did Moses know what science has only discovered in the last hundred years? How did Moses know these things 3,500 years ago? Could science be furnishing evidence that the Bible is the revealed Word of God?"

This theory has some very attractive features. It is an earnest attempt to take seriously both the Bible and scientific data. It avoids the scientific pitfalls of the literalism of "special creationism." It's true—there is a general correspondence between the sequences of Genesis 1 and what we currently know about cosmology. But there are problems, especially when we examine the details. Genesis speaks of the trees (day three) preceding marine organisms (day five), and birds (day five) preceding insects (day six). The fossil record clearly says otherwise. One very serious problem is the creation of the sun and stars on day four. How can there be evening and morning, and earth (a satellite of the sun) and vegetation before God makes the sun? Some very careful sidestepping can be done to reconcile these differences, but one is left with the nagging question: Why couldn't God have gotten it exactly right? Why does there have to be only a general correspondence? In addition, looking at similar past attempts to harmonize Genesis and science, it is evident that the approach continually requires us to reshape and fit Genesis to contemporary scientific explanations. Biblical interpretation becomes the servant of science, an undesirable and inappropriate relationship.

### *Literalist Theory*

Historically, major traditions within Christianity have held the view that the six days of Creation are literal days in which God accomplished his creative work, after which he rested. The reasoning behind this view is nicely summarized in the belief statement of the Creation Research Society:

> The Bible is the written Word of God, and because we believe it to be inspired thruout [sic], all of its assertions are historically and scientifically true in all of the original autographs. To the student of nature, this means that the account of origins in Genesis is a factual presentation of simple historical truths.

It is argued that the alternative to the literal reading is to consider the creation narratives as mythological or allegorical, and to do so would bring the rest of the Bible into question. The literalist view holds that the creative acts of God were unique, supernatural processes that occurred only a few thousand years ago. And since the earth and universe appear to be much older than this, it is sometimes argued that God created a "mature" earth, with the appearance of a much older age. Additional elements of this view include a rejection of evolutionary processes and, often, acceptance of *flood geology*—the belief that the Noahic flood accounts for the stratigraphy and presence of the fossil record. These are major facets of the beliefs of the *scientific creationists*, a group whose goal is to establish literalistic creationism as a scientific alternative to the conventional scientific views on origins.

The advantage of the literalist view is that it appears to allow a straightforward reading of the biblical text. The potential dangers of doing otherwise—of suggesting that the text is something other than a straightforward chronology—are presumed to be great enough to justify this view, even though it is incompatible with current scientific thought. Genesis is upheld as a historically and scientifically accurate record simply by taking the text at face value. Defenders of this view assert that scientific theories must not dictate to us how to interpret Scripture. *But:* the evidence from God's creation, from astronomy to zoology, indicates that the straightforward reading of the literalists could not be correct. Here is Blocher's view of this problem:

> The rejection of all the theories accepted by the scientists requires considerable bravado. It may be said that the work of many neo-catastrophist (= literalist) writers shows courage, not ignorance. Nevertheless, current opinions, built on the studies of thousands of research scientists who keep a very close eye on one another, continue to look very probable. Anyone rejecting them is taking an immense step. One must be absolutely sure of one's ground. . . . One must be sure that the text demands the literal interpretation.

How faithful is this interpretation to the text? One thing is certain; the literalist view requires the assumption that the major mode of God's activities in creation was supernatural—natural processes were suspended, and what did actually occur was outside the normal realm of cause and effect relationships over time. The text neither supports nor refutes this assumption. We would not question whether God could in fact carry off all of the creation events and processes using supernatural means; nor would we question whether he could do it using what we call "natural" means. He is God. But by loading origins down with supernaturalism, and flying into the face of modern science, the literalists have erected a very large stumbling block to belief in the God of the Bible. Their views have been and continue to be widely publicized, to the point that most people who are prompted to think about origins and Genesis associate creation with literalistic theories. Because of the vulnerability of this approach to a reasonable comparison with scientific findings, it becomes all too easy to dismiss the entire Bible and its claims. This is certainly ironic, because undoubtedly the prime reason for the literalistic theory has been to defend the authenticity of Scripture.

In summary, it seems clear that the reason for all of the above views, and the heated debates that have accompanied them, is the sequential, seven-day-week format of Genesis 1. Is this a literal week with literal days? If it is not, must we let our concerns to accomodate science dictate the interpretation of early Genesis? Or is there yet another interpretation, one that makes sense of the passage without violating exegetical principles? The following theory attempts to do this.

### *The Framework Theory*

According to the framework theory, the days of creation are arranged topically and not chronologically. It has a history that traces back to the church father Augustine, and currently enjoys the support of many theologians of strong evangelical persuasion. Instead of postulating that God dictated the seven-day sequence to Moses, it is easier to imagine Moses meditating on God's creation, on all of its components, and—inspired by the Holy Spirit—constructing an artistic account that makes clear how God is related to his creation and our place in it. He does this by constructing a literary framework that points to the theological framework of the Sabbath, hence the pattern of seven days. There is a logical sequence to the account; and if the actual chronology of origins agrees in some ways with the biblical sequence (as the day-age theory suggests), that was not the intent of the writer.

Here is the structure: Genesis 1:1 forms the thesis statement of creation: "In the beginning God created the heavens and the earth." Genesis 1:2 poses a problem: The earth was without form, it was empty, and it was dark (such that God's works could not be seen). Verses 3 through 29 then proceed to solve the problems, as shown in Table 1. The problem of formlessness is solved in days one through three, as God creates and separates the spatial environments into three static spheres. Then the problem of emptiness is solved in days four through six, as God creates and fills with moving forms the spheres formed in the first three days. The problem of darkness is the first to be dealt with; God calls light into being so that humankind may see God's works. At the same time, outer space is created. A minor discrepancy is the creation of land plants on day three; however, the plants were probably regarded as static elements, part of the habitat of the moving forms (indeed, Psalm 104 treats plants as "habitation," on a par with mountains). Before the sabbath is declared, Genesis 2:1 provides a summary statement that corresponds with the thesis statement that began the account: "Thus the heavens and the earth were completed in all their vast array."

*TABLE 1*

*Genesis 1 According to the Framework Theory*

| GOD FORMS (and separates) | GOD FILLS (and populates) |
|---|---|
| **Day 1:** Light and darkness separated (1:4) | **Day 4:** Sun, moon and stars (1:16) |
| **Day 2:** Atmosphere and oceans separated (1:7) | **Day 5:** Birds and aquatic life (1:20) |
| **Day 3:** Land and oceans separated; land plants created (1:9) | **Day 6:** Land animal life, human beings (1:24) |

Day seven is God's sabbath, a signal that with the creation narrative established, the rest of history as recorded in Scripture may commence. There is no evening and morning, for that "day" continues in time until now. The sabbath was clearly the dominating theme of this account; it is tied to the law in Exodus 20:11, and provides a structure for human activity that is God's wise provision: We shall work six days, and then enjoy a day of rest, when we also shall remember and commune with the Creator.

The careful construction and the everyday language point to this passage as a work of meditation, a work whose purpose was not to set forth for all ages a chronological history of God's creative work, but to picture God in relation to his creation in much the same way a painter might go about sketching a landscape and then filling it in. Is it history, does it speak of science? In a sense, yes; the events being portrayed really happened in time, and they very likely involved processes that we describe today as "natural processes." It is a bias of our modern mindset to look for precision and chronology and scientific information in such literature! This information is not to be found in the Bible, but in the creation, as we apply the methods of science to reach a coherent interpretation of origins.

The framework theory, then, fosters a compatible relationship between the Bible and science. This is in contrast to the difficulties of other interpretive theories in conforming to the testimony from both God's word and God's world at the same time. Blocher ends his discussion of the framework theory with these words:

> What are we to conclude? The theological treasures of the framework of the Genesis days come most clearly to light by means of the "literary" (= Framework) interpretation. The writer has given us a masterly elaboration of a fitting, restrained anthropomorphic vision, in order to convey a whole complex of deeply meditated ideas. Of that we have no doubt; though whether it is the content or the form that calls forth the greater admiration, we cannot tell.

The major objection to the framework interpretation, as articulated by biologist Pattle Pun, is the apparent separation of theological and historical dimensions to this passage. Pun refers to the existence of eleven historical narratives in the first thirty-seven chapters of Genesis, each of which starts with the phrase, "These are the names (generations, descendents) of . . ." (Yet the first appearance of this form is in Genesis 2:4, where the second and summarizing creation account begins). Pun summarizes:

> While Blocher's framework hypothesis is attractive for its resolution of some of the apparent conflicts . . . it remains unclear at what point one can draw the boundary line between an allegorical account, where only the spiritual meaning prevails, and a historical-theological account, where both what actually transpired and its spiritual meaning are significant.

In my view, the crucial question is whether the writer of Genesis 1 was presenting a chronological account or a topical, meditative account. Clearly, Christians with a high view of both Scripture and science do not yet agree on the answer. What do you think?

## Models for Relating Scripture and Science

Now that we have considered several alternative approaches to Genesis 1, it is time to return to the larger question of relating Scripture to science. We now have a more solid basis for describing and evaluating several interactive models, and we will give special weight to their impact on the biological sciences.

### *Concordism*

The fundamental distinction of concordism is the belief that Scripture contains vital information about the natural world that can supplement the information gathered by the direct study of nature, and that these two sources of information will harmonize when they are properly understood. In other words, there are gaps in both the biblical and the scientific record, and a thorough understanding, especially of origins, will only come from study of both sources of data. (Note that this presupposition will have a strong effect on interpretation of Scripture.) This model is very attractive, and has many followers. Both Scripture and science are taken seriously, and frequently the results bring about a reduction in the tensions between these two. Proponents of this model have a high respect for scholarship, and their publications are nonaggressive in tone.

How do concordists treat biology (primarily, evolution)? They are much more receptive to geological and cosmological findings than to evolutionary theory. Although they accept the extreme antiquity of the earth and the fossil record as authentic, there is often a reluctance to ascribe all of the development of plants and animals over time to evolution. They focus on gaps in the fossil record (lack of Precambrian fossils, for example) and on the lack of transitional forms (such as a transition from invertebrates to fish), and call into question a completely evolutionary explanation. Instead of evolution, the concordists often suggest that God may have created living forms at crucial intervals in the history of life (*progressive creation*), and then allowed evolutionary mechanisms to produce closely related species over time.

This attention to the weak points in the fossil record is certainly in the best traditions of science, but in the end it reveals a serious problem with the Concordistic approach. It leads to the "God of the gaps" problem. When we make God responsible for those things that we currently cannot explain, the gaps in our explanations, we open up a line of reasoning that leads to a denial of God as soon as a natural explanation is found. As we saw in chapter 2, the Bible makes it clear that God is responsible for all of what we think of as natural processes—that through Creational Law the whole creation exists in a covenant, lawful response to God. A God who carries out his purposes by periodically intervening in the natural order comes very close to deism. And concordism leads to frequent reinterpretation of Scripture in order to accomodate the latest scientific findings—an undesirable and risky procedure.

### *Substitutionism*

Substitutionism has been called various things, but its fundamental presupposition is the view that Scripture contains scientific truth, and because the Bible is God's inerrant and authoritative word, Bible science is more trustworthy than conventional science. We are to substitute the more trustworthy science of the Bible (creation science) for the naturalistic interpretations of scientists, and this is especially true for understanding origins. In short, substitutionists believe that the strongly naturalistic worldview approach of most proponents of evolution introduces serious bias into their scientific work.

The substitutionists confront biology with a totally different paradigm for the origin of species and for the adaptations of living things to their environment. They are uncompromising in their criticism of evolutionary biology, and have raised a great deal of concern from the scientific establishment because of their attempts to establish equal treatment of their views with evolutionary views in the public schools. The choice—either creation or evolution—has been broadly debated in scores of recent books (we will examine some aspects of this controversy in chapter 7).

With a few exceptions, most of the critics of evolution who write from this perspective are either physical scientists or are in some other field of endeavor. This is not to say that the majority is necessarily right; rather, I would contend that most Christians who hold substitutionist views because of the creation/evolution issue do not have a thorough understanding of the biological basis for evolution. But their basic presupposition about the Bible and science forces them into a posture of confrontation with evolutionary science. This confrontation has had some positive effects, in that it has pointed out the strong worldview thrust of evolution (we would call this *evolutionism*) and has tended to curb some of the more excessive claims of scientists in the area of origins. But it has also forced many Christian biologists to spend much time defending their views from attacks by fellow Christians, and has placed a major stumbling block to faith in the way of those who might give biblical truth a hearing if it weren't identified with creation science. Blocher has some interesting things to say here:

> In order to have the right to challenge beliefs held by the vast majority, one must show that their presuppositions lead to misunderstanding the facts. If possible, a better general theory must be elaborated. The exponents of anti-scientism (= substitutionists) do their best to do so. Have they succeeded? Observers who stand with them in the area of faith are dubious on this point: they . . . underestimate the value of the scientific consensus reached in the world. They forget that the mutual criticisms of the specialists who are often each other's rivals protect them partly from unwarranted extrapolations. The agreement of thousands of researchers is reached neither by chance nor by conspiracy! In our eyes, the opponents of established opinions, the kamikaze pilots of the academic world, show their lack of weight at two decisive points: when they minimize the value of identical conclusions reached independently by divergent methods. . . . and when, without permitting discussion, they attribute to Genesis a meaning which other readers do not find there and which they themselves justify only on the a priori ground of literalism.

### *Compartmentalism*

The basic presupposition of compartmentalism is the conviction that science and religion deal with entirely different realms. There is the realm of faith, and the realm of science, and they must be kept apart. The Bible, the compartmentalist would say, is not a handbook of science; further, there is no common ground on which the Bible and science could meet. In essence, the problems that we have been examining in this chapter should never come up. Early Genesis is seen as mythological, or a series of parables, teaching us theological truths but lacking historical reference points or any information about the natural world. This approach avoids conflict between science and Scripture by defining the problem away. Evolution presents no problems to the compartmentalist unless it is extended into a worldview that excludes the possibility of Christian faith.

The crucial problem with this model is its limited view of biblical authority (which accompanies an unsatisfactory view of inspiration). The biblical worldview is the entire world, not just the world of faith. The Bible speaks of God's actions in history, of God's creation of all realms. Science is a study of God's world, and he has a number of important things to say about that world. And biblical faith rests on facts and a God who has acted in history. Del Ratzsch comments on this model:

> It is not really clear how the respective realms are to be divided. In fact, religious statements and scientific statements are often about some of the same subjects. For instance, we can make biological statements about trees, but we also have to say that those same trees are creations of God. . . . We can make scientific statements about humans, but we must also say that those same humans are created in God's image. A strict separation of the items of creation into those wholly subject to science and those wholly subject to religion does not then seem successful.

### *Complementarism*

The complementarist model assumes that both biblical truth and scientific knowledge are needed for a balanced view of origins and the natural world. They are not competing views, neither are they completely separate; they complement each other. They offer different kinds of explanations, because they ask different questions, employ different methodologies, and have different purposes. They are, in a sense, different maps for the same landscape.

When, for example, origins is the subject, the biblical message about origins consists of a clear statement that God is the author of all of the universe, that he created with wisdom and purpose, and that he considered his creation to be of great value. This is not a scientific statement about origins, but it gives us information that science is not capable of discovering—hence it is complementary to the scientific view. On the other hand, scientists may investigate origins and offer their explanations of the data in the form of explanations from within the natural world. If they are properly expressed, those explanations make no attempt to argue first causes, to assign values, or to proclaim complete autonomy of the natural world. There are limitations, therefore, to both approaches, but that should not surprise us.

It might be said that this model makes too many concessions to science. On the contrary, complementarists would argue that their approach simply recognizes the limitations of both fields and allows them freedom to generate their complementary explanations of the created world. How does this model deal with biological science? Since the Bible has no intention of establishing biological concepts, biologists are free to examine the data and reach their conclusions independent of any constraints to harmonize with biblical data or to fit into a literalistic biblical framework. They must be careful, however, to recognize the limits of their science, and too often this has not been done. Evolution is frequently elevated to worldview status, as we saw in the last chapter.

### Forming Your Own Opinion

These are some of the views that different Christians have put forth in order to provide a framework for understanding how Scripture and science relate. If I were to advise someone starting out on a biological pilgrimage, my advice would be (1) look thoroughly into these things and form an opinion based on your best judgment; (2) don't hold on to that opinion very tightly; and (3) be slow to condemn those who hold different views.

My own views have changed quite a bit over the years, and I suspect that will be true of most Christians who, like me, enter biology and only gradually come to appreciate the depth of the subject and the complexities of its connections to Christian thought. I continue to be impressed with the great privilege I have to be both a Christian and a scientist. I don't know how these two areas interact in my mind (in concordistic, substitutionistic, compartmentalistic, or complementaristic ways?). But I know that I am a better scientist because of my Christian faith and that my scientific work has helped me to live a fuller Christian life.

### Summary

Genesis 1 leads directly into origins issues that Christians in biology must confront. Strong differences exist between Christians in the interpretation of Genesis 1; doctrinal issues are at stake, but several available interpretations preserve crucial Christian doctrines. A major contention surrounds whether the findings of science can be brought to bear on this passage. It can be argued that scientific knowledge is part of God's truth, even though we grasp that truth fallibly. Therefore science—as well as history and language study—can be used in interpreting early Genesis, as long as the limitations of science are kept in perspective.

Henri Blocher suggests that this passage represents a composite of different literary modes, having elements of poetry and prose. Symbolic numbers are heavily used. The language is that of everyday experience, nontechnical and reflective of the worldview of ancient Hebrew culture. It was probably written during the wilderness wanderings of Israel, and was intended to teach the Israelites that their covenent God was also the creator of all things. There is a definite correspondence between the creation week and the establishment of the Sabbath within a seven-day week.

Four interpretations of Genesis 1 are presented. The *reconstruction theory* postulates an original creation in Genesis 1:1, which was destroyed when Satan fell and evil entered the world. The six days of creation are then six literal days during which God reconstructed things. The *day-age theory* holds that the creation "days" are long, possibly overlapping periods of time during which God carried out creation. The use of natural forces by God is considered the normal mode, with occasional supernatural creations at critical junctures. The *literalist theory* argues that creation took place as a series of supernatural creative acts of God in six literal days, and that the earth is therefore much younger than scientists believe. The *framework theory* suggests that the days of creation are arranged according to topics, not chronologically. There is a literary framework that shows God in the first three days forming and separating spatial environments, and in the second three days respectively populating the spatial environments with moving forms. The sabbath of the seventh day is the dominating theme of the passage.

Each of these four theories is examined with respect to the interpretive problems it presents. Major emphasis is placed on the degree to which a theory is compatible with well established scientific knowledge. The different interpretations of Genesis 1 lead to examination of a major problem introduced in chapter 3: How do science and Scripture relate?

Four interpretive frameworks for relating science and Scripture are presented and evaluated on the basis of compatibility with biology. *Concordism* holds that the Bible contains information about the natural world that will harmonize with the information coming from scientific studies. *Substitutionism* is the view that the Bible contains scientific truth which, if it does not agree with modern science, is to be substituted for that science. *Compartmentalism* suggests that science and Scripture deal with entirely different realms that should be kept apart. *Complementarism* holds that science and Scripture present different but complementary ways of looking at the natural world.

The chapter ends with some advice on how to form an opinion about these different approaches to science-faith problems. It is suggested that in spite of the problems, approaching the world from the combined perspective of science and Christian faith is still a great privilege.

*Chapter 6*

# THE ORIGIN OF LIFE

> And, therefore, gentlemen, I could point to that liquid and say to you, I have taken my drop of water from the immensity of creation, and I have taken it full of the elements appropriated to the development of microscopic organisms. And I wait, I watch, I question it!—begging it to recommence for me the beautiful spectacle of the first creation. But it is dumb, dumb since these experiments were begun several years ago; it is dumb because I have kept from it the only thing man does not know how to produce: from the germs which float in the air, from Life, for Life is a germ and a germ is Life. Never will the doctrine of spontaneous generation recover from the mortal blow of this simple experiment
>
> LOUIS PASTEUR

## Ultimate Origins

Both the Bible and science agree that at some time in the very distant past, life on earth did not exist. But our imagination balks at any attempt to picture how life could have gotten started. For this reason, the Christian view comes as something of a relief: *God created life.* That is only the beginning of the matter, so to speak. To quote the title of a fine book by chemist Robert Fischer, *God Did It, but How?*

In trying to answer how life might have begun, scientists are looking into the past and offering possible explanations for the origin of life as a result of biochemical evolution. Is it then possible to bring these two viewpoints together, and offer this as an example of the complementarity approach? Or are there other issues involved, so that Christians are not able simply to endorse what scientists are saying but must insist on something different?

### *Spontaneous Generation*

Louis Pasteur, that great French biologist of the nineteenth century, studied microorganisms and verified that such processes as the souring of milk, the fermentation of grape juice, and the decay of meat were caused by microbial activity. He felt certain that the microorganisms promoting these changes were already present in the materials, rather than being generated spontaneously by the organic matter. At the time—1864—spontaneous generation was an unsolved problem, and it was often cast in the mold of the origin-of-life question. Pasteur conducted a famous series of experiments with his "swan-neck flasks," flasks that were in contact with outside air through a long, curved neck. By filling the flasks with a fermentable fluid and then boiling the contents, Pasteur was able to test the ability of the fluid to develop microbes. His words describing the results of the experiments before a distinguished audience at the Sorbonne are found in the quote that opened this chapter.

As all good biology texts will tell you, Pasteur's crucial experiment did indeed bury the theory of spontaneous generation. Life comes from life—biogenesis. But while Pasteur was boiling his flasks, Charles Darwin was publishing *The Origin of Species* and providing some new answers to the questions about adaptations and the origin of new species. We saw in chapter 4 how most of the scientific world quickly "converted" to an evolutionary viewpoint following Darwin's lead. The same scientists were fully aware of Pasteur's conclusions, and soon encountered a dilemma: Darwinian evolution requires living, reproducing organisms and then proceeds to show how they evolve, all of which may be neatly incorporated into a naturalistic worldview. But if life could not appear spontaneously, then the only alternative was creation by God, and that was unacceptable. A believable theory about how life could appear *by chance*, then, would be a nice complement to Darwinism and a necessary element for keeping God out of the picture.

In a brief article on the origin of life, Harvard biochemist and Nobel laureate George Wald comments on the dilemma facing those who wanted to maintain a naturalistic explanation for life's origin:

> We tell this story (of Pasteur's experiments) to beginning students of biology as though it represents a triumph of reason over mysticism. In fact it is very nearly the opposite. The reasonable view was to believe in spontaneous generation; the only alternative, to believe in a single, primary act of supernatural creation. There is no third position.

### *An Inappropriate Choice*

This topic clearly has metaphysical (worldview) dimensions. The desire to find a naturalistic explanation was present long before any progress was made in research and reasoning on origin-of-life questions—well into the twentieth century. The thrust of this issue is the choice between life being created by God, which is considered to be a religious matter, and life originating spontaneously from inorganic matter, a scientific matter.

This choice is inappropriate for two good reasons. First, it implies that God would only be involved if life originated by supernatural or miraculous means. As we saw in chapter 2, this is an unacceptable restriction on God's activities in his world. All of creation is an expression of his being, upheld by the power of his Word—he governs the universe. In its operation in space and time, the natural world obeys God's Creational Law as a covenant response to him.

Second, the evidence for the spontaneous origin of life is remarkably sketchy. Astronomer Robert Jastrow says of this alternative:

> The first theory places the question of the origin of life beyond the realm of scientific inquiry. It is a statement of faith in the power of a Supreme Being not subject to the laws of science.
>
> The second theory is *also an act of faith* [my emphasis]. The act of faith consists in assuming that the scientific view of the origin of life is correct, without having concrete evidence to support that belief.

Unfortunately, many textbooks are less than candid on this point. We will proceed by first summarizing what some recently published biology texts present as an explanation of the origin of life, and then take a closer look at some of the evidence before rendering an opinion on what a Christian view on all of this should be.

## The View from the Texts

Most texts begin a discussion on the origin of life by giving credit to the thinking of a Russian biochemist, A. I. Oparin, whose book *Origin of Life on Earth* first appeared in an English translation in 1938. Oparin's work was followed by many others (including J. B. S. Haldane, Harold Urey, and Sidney Fox), and gradually there developed a consensus of what could now be called the modern theory of chemical evolution. A sure sign that such a consensus exists is the fact that almost all biology texts include it in some form in their discussion of evolution.

### *Stages in Chemical Evolution*

The modern explanation involves five stages, illustrated in Table 2.

- Stage 1. The first stage begins with the formation of a favorable environment. This consisted of a primitive atmosphere made up of a mixture of gases: water vapor, nitrogen, carbon dioxide, methane, ammonia, and hydrogen, but no oxygen. Thus it was a reducing rather than an oxidizing atmosphere. By this time the earth had cooled, water had condensed, and the primitive ocean had appeared. An abundant source of energy was present in the form of ultraviolet radiation.
- Stage 2. Organic monomers—simple organic compounds like sugars, amino acids, and fatty acids—were formed as a

## TABLE 2

*Stages in the Origin of Life**

| **Stage 1** Early earth atmosphere | **Stage 2** Hot dilute soup | **Stage 3** Widescale polymerization | **Stage 4** protocells | **Stage 5** True cells |
|---|---|---|---|---|
| Water | | | | |
| Hydrogen | | | | |
| Methane | | | | |
| | Fatty acids → | Lipids → | Membranes ↘ | ○ |
| Carbon monoxide | Amino acids → | Peptides → | Proteins → | ○ |
| Carbon dioxide | Sugars → | Carbohydrates ↗ | | ○ |
| Ammonia | Purines and pyrimidines → | Polynucleotides → | RNA/DNA ↗ | ○ |
| Nitrogen | | | | |

* Adapted from Thaxton *et al.*, Figure 2–1.

result of reactions of the compounds in the atmosphere and the waters with solar radiation; some picture the ocean as a vast, dilute soup of organic compounds—the "primordial soup."

- Stage 3. Extensive polymerization of the monomers occurred, stimulated by the continuous energy input and made immune to degradation by the lack of oxygen. Concentration of the "soup" in isolated pools or on clay particles would have speeded up the polymerization process. These polymers may have been proteinaeous, or nucleic acids, or perhaps both.
- Stage 4. Formation of protocells, as polymers became bound by membranes and were separated from the external medium. These protocells may have been coacervate

droplets (Oparin) or proteinoid microspheres (Fox), or something else, but they were constructed of biochemical polymers, separated from the medium, and possessed some internal structure.

- Stage 5. The final stage was the formation of the first primitive cells, capable of both reproduction and metabolism. These were undoubtedly heterotrophs, which could metabolize the organic matter that had been forming for vast numbers of years.

### *Reliability of the Model*

Textbook presentations of this sequence range in tone from the very positive to the highly tentative. An example of the former: "With further cooling, water vapor condensed to form the oceans, and the gases trapped within the oceans underwent a chemical evolution, an increase in complexity that eventually produced the first life molecules and then cells." Another text uses more appropriate wording: "a number of scientists have wondered . . . it has been postulated . . . it is believed . . . if, in fact, cells evolved by chemical evolution . . ."

A similar range of approaches can be found in the cluster of books recently published in answer to the scientific creationist attack on the teaching of evolution in public schools. Concerning the chemical evolution of life, one states, "a combination of geochemical evidence and laboratory experiment shows that such evolution is not only plausible but almost undeniable," and "It's likely that simple forms of life will be synthesized in the laboratory within the next ten years." And from another such book: "It can be stated with some confidence that, in outline, we know how life evolved from its earliest beginnings." Oddly, though, earlier in the same book the author shows more candor: "As we move along through the stages of the putative evolution of life, increasingly we enter the realm of the hypothetical." Clearly, where the author's design is to debate the issues, as in the latter books, or to instruct the neophyte biologist in origins questions, there are strong tendencies to represent origin-of-life research and the conclusions from that research in a very positive (shall we say, naive positivistic?) light.

It is a refreshing commentary on the self-correcting character of science that Francis Crick, a Nobel laureate who can speak with some authority, can wade into this controversy and issue a counter opinion: "The origin of life appears at the moment to be almost a miracle, so many are the conditions which would have had to have been satisfactory to get it going." Crick's idea is that life was introduced from outer space by some higher intelligence, perhaps on the wings of a missile ("directed panspermia," as he calls it). He does say, however, that in this question of life's origin we find" too much speculation running after too few facts." This charge is sufficiently important to pursue further—especially as it might help us to decide what the stakes are in this issue.

## Major Problems with Chemical Evolution

### *Primitive Reducing Atmosphere*

What kind of primitive atmosphere would be conducive to the formation and accumulation of organic monomers? On this point there is virtually unanimous agreement: a *reducing atmosphere*, one where oxygen was absent and compounds like methane, hydrogen, and ammonia were present. Why? Because oxygen, even at low concentrations, would inhibit the formation of precursor organic compounds and, if they were already present, would promote their decomposition. These two reasons are cited as the strongest evidence in favor of the reducing atmosphere hypothesis by chemists Sidney Fox and Klaus Dose, authors of *Molecular Evolution and the Origins of Life*. Can you see the circular reasoning here? The major hypothesis: Chemical evolution led to the origin of life. The argument: Chemical evolution could not occur in the presence of oxygen. The conclusion: Therefore oxygen must not have been present (the conclusion is actually part of the hypothesis being examined!).

The most recent work suggests that the early atmosphere contained the same mixture of gases that are found in present-day outgassing of volcanoes: water, carbon dioxide, nitrogen, and small amounts of hydrogen. This mixture is approximately neutral (not oxidizing or reducing). The vital question then is, was oxygen present? The source of oxygen, in the absence of photosynthesis, would be photodissociation of water (the separation of water into hydrogen and oxygen using light energy); and it has been suggested that since initially there would not have been an ozone shield, the high amounts of ultraviolet light would have led to significant oxygen production (and then to the presence of ozone).

The only direct source of evidence on the presence of oxygen is the types of minerals found in rocks (reduced minerals should be present in the most ancient sedimentary rocks, dated as old as 3.4 billion years before present), and this evidence is no help, as the following quote indicates: "In general, we find no evidence in the sedimentary distribution of carbon, sulfur, uranium, or iron that an oxygen-free atmosphere has existed at any time during the span of geological history recorded in well-preserved sedimentary rock." The conclusion of the matter is that early "classical" origin of life experiments showing the production of organic monomers in a reducing "atmosphere" need to be reassessed. It is by no means established that the early atmosphere lacked oxygen.

### *Chemical Evolution of Life's Precursor Compounds*

The experimental approach to chemical evolution consists of attempts to show the production of biologically important monomers and polymers under conditions thought to simulate those of the early earth. These experiments have enjoyed a great deal of success and—to be sure—prominence for the researchers. Several sugars, all five of the nitrogenous bases of nucleic acids, and almost all of the amino acids have been synthesized in what have been called *prebiotic simulation* experiments. However, the conditions for the experiments are carefully defined and artificially simplified to the point where their applicability to chemical evolution may be seriously challenged.

For example, the reactants of the experiments are carefully regulated and the products are condensed in traps that remove them from the reaction medium. Different classes of compounds are kept from contact with each other, thus preventing cross-reactions that would tend to produce complex and insoluble messes. This is an especially critical issue when imagining the creation of polymers of, say, amino acids (proteins), combinations of bases, sugars and phosphate (nucleic acids), and sugars (carbohydrates). These and other experimental conditions create a highly significant *investigator interference factor.* At the end of a review of abiotic experiments, J. Brooks and G. Shaw state: "These experiments . . . claim abiotic synthesis for what has in fact been produced and designed by highly intelligent and very much biotic man."

### *Synthesis of Cells*

The transition from biopolymers to the first living cell is at present not even imaginable. As information on the complexity of structure and regulation accumulates from molecular biology, the prospects of "creating life" in the test tube, as it is often expressed, recedes farther and farther into the mists of the future. The crude "coacervates" and "microspheres" of the experimentalists are mere images of the real thing, having something of the form and almost nothing of the substance of a living cell.

Harvard biologist Ernst Mayr points out a major problem:

Curiously, the findings of molecular biology have complicated the task of explanation rather than simplified it. Polypeptide chains (proteins), even in the simplest organisms, are assembled from amino acids under the guidance of a nucleic acid genetic program. Indeed, there is now such a complete "symbiosis" between nucleic acids and proteins that it is difficult to imagine either being able to function without the other. How then could the first proteins have been assembled and replicated without nucleic acids, and how could nucleic acids have originated and been maintained in the primeval "organic soup" if they had no other meaning than to control the assembly of proteins.

More recent research has uncovered at least a partial answer to Mayr's question: Some forms of RNA are able to act as enzymes, catalyzing the reactions of other nucleic acids. This is only one of many problems in the path of trying to understand how the first cell might have evolved, and a suitable summary statement says it clearly: "the macromolecule-to-cell transition is a jump of fantastic dimensions . . . The available facts do not provide a basis for postulating that cells arose on this planet."

### *Life Goes On*

But life did appear on earth, a long time ago. The fossil record indicates that living cells occurred at least 3 billion years before present. These first cells were apparently procaryotic cells, but nothing is known of their internal structure or of their nutritional mode. Once the first cell appeared, evolution by natural selection and mutation became a possibility. The accounts of early life that appear in the textbooks suggest that the first cells were heterotrophic, using the contents of the primordial soup for fermentative growth. Autotrophic procaryotes (chemosynthetic bacteria and blue-green algae) appeared later, and the oxygen that was given off by photosynthetic microorganisms led to the evolution of aerobic organisms. There is no evidence suggesting that this scenario is correct; it happens to be the most reasonable based on what we know about current living organisms and the very limited fossil record of early life.

### Operation Science and Origin Science

Charles Thaxton, Walter Bradley, and Roger Olsen teamed up to write *The Mystery of Life's Origin*, from which I have drawn heavily in discussing origin of life issues. These authors have pointed out an interesting distinction between operation science and origin science. It is one that we need to examine carefully.

*Operation science* is the science most of us are familiar with—the science that attempts to explain the recurring phenomena of the natural world. These phenomena occur repeatedly, and so can be tested by experimentation and by repeated observation. Hypotheses to explain them are therefore subject to the test of falsification. The flow of energy through ecosystems, the action of hormones on receptor cells, the transmission of an action potential down a nerve axon, the transport of water through xylem tissues—all of these are details that are susceptible to explanation by applying the experimental and observational approaches of modern science. Much of biology consists of operational science. Thaxton *et al.* suggest that we should not resort to supernatural intervention by God in order to account for any of these phenomena. If we don't understand them, we should get to work and use the methods and reasoning ability already available to us in order to generate an explanation, and not suggest that God does it in some kind of miraculous way. To bring God into the explanation in this way would be to fall prey to the God-of-the-gaps problem we mentioned earlier.

On the other hand, *origin science* attempts to understand unique events in the history of the cosmos. Since these events are nonrecurring by definition, we cannot test them experimentally. Whereas operation science focuses on a class of events that repeat over time, origin science deals with events that have occurred only once. The origin of life, the movement of continental plates, the formation of fossils—indeed, the formation of species of plants and animals and of their adaptations—all of these would fall under the scope of origin science. It is the legitimate task of science to develop explanations of these phenomena, even though we are not able to apply the method of experimentation.

However (the argument continues), true falsification of a theory in origin science is not possible. There is simply no way to test experimentally something that happened so long ago. The most we can hope for is to show that a particular explanation is either probable or improbable, by using the evidence we have and testing our explanation in the experimental arena of operation science. Because of this limitation, Thaxton *et al.* caution strongly against removing *divine intervention* from the events covered by origin science. There is no justification for assuming that they happened by means of "natural processes" (in terms of our earlier view of the structure of science, such an assumption is a *shaping principle*, a presupposition one brings to the scientific enterprise). As a conclusion to their work on the origin of life, Thaxton *et al.* argue that special creation by God is a more believable explanation of how life originated than the explanations involving chemical evolution.

### *Evaluation*

This distinction between operation science, which deals with repeatable phenomena, and origin science, which is retrospective, implies that there are in fact two kinds of science. If this is true, the consequences are important. Operation science is the more reliable, whereas origin science has limited ability to generate scientific explanations. I would like to suggest that the distinction between these two kinds of science is not as strong as Thaxton *et al.* suggest.

First of all, they place strong reliance on the process of *falsification* for the experimental approach of operation science. The implication is that if a prediction of a theory is shown to be false, the theory is also false. Conversely, if the prediction turns out to be true, the theory is confirmed. But this is part of an outdated view of the scientific method. Contemporary philosophers of science point out that our theories cannot be proven either true or false. The falsifiability principle implies that theories will automatically generate their predictions, which we can then test experimentally. However, the predictions are actually made by the scientist; in the process, important background theories (shaping principles) about the prediction and the apparatus used in performing the experiment are assumed to be true. Since no theory can be proven true, the reasoning used in generating and measuring the effects of the prediction might also not be true, and so the theory under test cannot be proven false. Operation science therefore is not on as firm a footing as Thaxton *et al.* would suggest.

The distinction between operation science and origin science also implies that the latter is less capable of generating a believable theory about events in the natural world because it lacks recourse to experimental testing. In fact, in the case of a historical event—take, for example, the splitting of one species into two following the appearance of a geographical barrier between populations—an experiment actually has taken place. It is an "experiment of nature," as Ernst Mayr calls it, and such natural experiments are capable of being studied and explained by use of the *observational-comparative method*. This is the method most heavily used in geology, meteorology, and astronomy, as well as in evolutionary, ecological, and behavioral biology. Mayr argues that this method is every bit as scientific and as capable of generating plausible explanations as the experimental method. The task of the observational-comparative method is to theorize about the conditions under which a given experiment of nature took place, which then leads to the development of an explanation. The explanation can earn a high degree of plausibility if the evidence for the experiment of nature is sufficiently available. Mayr points out that although this method relies heavily on observations rather than experiments, the experimental method must also make observations—of the experimental results. The only real difference between the two methods is that the experimentalist can choose the conditions of the experiment, and repeat it, while the observationalist cannot.

Let's consider the observationalist's approach to the origin of life. It would be unrealistic if we assumed that anyone would approach the question without some prior hypotheses. Clearly, the key hypothesis is that life originated spontaneously at some distant time in the past. What are the alternatives? We could have Crick's *panspermia*, or *special creation* by God, both of which might suggest that life's origin on earth is outside the limits of science. As I see it, the methods available for investigating the origin of life are (1) examination of evidence from the past, in the form of geological evidence; (2) examination of evidence from the present, in the form of present living things and their structure as a function of the information encoded in DNA (remember, living things have a history); (3) the involvement of operation science factors—mechanisms that can be extrapolated to the past; and (4) the development of an analogical model by experiments attempting to reconstruct what might have happened (synthesis of monomers, polymerization into macromolecules, and so forth).

The data from the geological record (method 1) do not support or refute any of the hypotheses. There is no evidence in the rocks for the organic soup that is supposed to have been in the primordial ocean; there is not clear evidence from the minerals in sedimentary rocks for the existence of a reducing atmosphere. Fossil evidence basically shows that life goes back at least 3 billion years, but so far no protocells have been found. The data from present-day life (method 2) provides us with information about what would have to be present in order to have a living cell, and suggests to us in a number of ways (such as the universal genetic code) that living things are related and may have come from a common source in the past. It does not support (or refute) any of the hypotheses, however.

Only methods 3 and 4 remain if the hypothesis of spontaneous generation of life is being examined. Both of these methods involve the reconstruction of conditions under which the origin of life might have occurred, and since it happened so long ago and geological evidence is lacking, it seems certain that any reconstruction will have a fairly low degree of reliability. As we have seen, the development of reconstructive models is fraught with difficulties; it is not hard to find weaknesses in this kind of research, as Thaxton *et al.* have shown clearly.

At this point, it might be instructive to examine the origin-of-life question from the perspective of the four models discussed in chapter 5 for relating Scripture and science, and then make a final evaluation.

## The View from the Theistic Models

### *Concordism*

According to a recent account of origins from this perspective, noted in John Weister's *The Genesis Connection*, Genesis 1:11 and 12 suggest that God used materials from the earth to create vegetative life, which is presumed to be the first form of life on earth. The wording is important: "Then God said, let the land produced vegetation. . . . The land produced vegetation . . ." Accordingly, the language suggests that God could have used preexisting materials and energy to bring about the origin of life, and the process might be chemical evolution.

Another concordistic account suggests that God deliberately engineered the first life, as well as succeeding organisms at different times in history: "No intelligent Creator would leave matters to chance; on the contrary, his purpose would be to realize, in plan and in practice, his ideas." In summary, concordism might argue that life originated long ago, as the fossil record indicates, but that God may or may not have employed natural mechanisms in creating it. Under no circumstances could life's origin be seen as the outcome of chance and "natural" mechanistic forces; the origin of life was not "spontaneous," in the true sense of the word.

*Substitutionism*

From this perspective, there is no point in attempting to show how life could have originated as a result of the interaction of chemicals and energy in a primitive atmosphere, long ago. The Bible indicates that God created living things relatively recently, and the mechanism of "special creation" is clearly miraculous, involving a suspension of the natural laws. The complexity of living cells, the lack of geological evidence, and the flaws in design of chemical evolution experiments are all cited as arguments against the spontaneous generation of life and evidence in support of special creation. The first life was land vegetation, on the third day of creation, according to this view. Life, from origin to its present developmental stages, is the outcome of deliberate design and special creation by God.

*Compartmentalism and Complementarism*

Neither of these views has a serious problem with the attempts to explain life's origin as the outcome of chemical evolution. The Bible simply does not supply information about how God brought about the origin of life. As long as scientists confine their reasoning to the realm of the internal operation of the cosmos, the complementarist viewpoint would find no fault with origins research. Only when the scientific explanation asserts that God had nothing to do with life's origin, that it was a matter of pure chance that life appeared at all, is there is a problem. Science in this case has become scientism, reaching well beyond its limits in order to make judgments that are clearly an expression of an antitheistic worldview.

### Where Does That Leave Us?

The key hypothesis in origin-of-life issues is that life originated spontaneously at some time in the distant past, as a result of the interaction of natural mechanisms. We have briefly considered the evidence in support of this hypothesis, and found it wanting. Yet it is the only hypothesis that will allow us to attempt to explain what might have happened, while remaining within the limitations of science. Perhaps as more work is done, the reconstruction will improve in its plausibility.

Let's suppose some scientist actually "creates" a living form in the laboratory along the lines of the origin-of-life scenarios. Two responses are appropriate: First, a highly intelligent being designed and carried out the experiment—and this is not a trivial observation. Second, the result would be only one of an almost infinite number of variations on the theme, with no way of knowing which one is right. Undoubtedly, if and when a living form is created in the laboratory, many who hold naturalistic worldviews will celebrate the apparent triumph of reason over superstition. Indeed, it is hard to escape the conviction that the desire to find a naturalistic explanation for the origin of life is the main driving force behind much of the research in this area.

Shall we object to this origin-of-life work, because of our theistic beliefs? To do so would suggest that something vital is at stake—that if the origin of life by spontaneous generation became well established, we would lose yet another battle for keeping God involved with his world. This is the "God of the gaps" trap, and it should be avoided. But the fact is, the present evidence for the spontaneous origin of life is not strong enough to establish a plausible reconstruction, and it is certainly not strong enough to discourage belief in other options, such as the special creation of life by God. Thaxton, Bradley, and Olsen have looked at the evidence and have concluded that the engineering feat that must have been involved in constructing the highly organized information encoded in DNA, and expressed in the protein structures that enclose and maintain the cell, is far too complex to be explained by the interplay of natural forces. They have opted for special creation, by a miraculous process. In other words, they feel that the boundary between the nonliving and the living is a major discontinuity, a change so profound that only a special act of God could bridge it.

But their conclusion is not the only theistic option. The fundamental conclusion of faith is that God created life, and our creaturely limitations force us to conclude that we do not know how. As we saw in chapter 2, God's providential care for his creation is the outcome of Creational Laws; our descriptions of how matter and energy behave—"natural laws"—are at best dim and incomplete reflections of the operation of these Creational Laws. It is entirely reasonable to suggest that the origin of life came about as a result of what we might call natural mechanisms, and at the same time maintain that God was acting to carry out his purposes. Biologist Dave Wilcox has stated the case well: "Anyone who is a fully biblical theist must consider ordinary processes controlled by natural law to be as completely and deliberately the wonderful acts of God as any miracle, equally contingent upon his free and unhindered will."

## SUMMARY

The Christian view that God created life must be considered in the light of work by scientists to probe the origin of life in the distant past. Pasteur's demolition of the spontaneous generation theory created a problem for people who wanted to maintain a naturalistic explanation for life's origin, in that it seemed to offer special creation by God as the only alternative. The choice seems to be a faith belief in God as creator of life, versus a scientific belief in the spontaneous generation of life from inanimate matter. This is a poor choice, because God would be involved no matter how life started, and because the scientific evidence for spontaneous generation also requires considerable faith.

Most texts present the origin of life as occurring in several stages, and the presentations range in tone from highly dogmatic to somewhat tentative depending on the text. In general, where the issue is being debated or new biologists are being instructed, the tone is highly confident and dogmatic.

Major problems with the scientific views center on: (1) the question of oxygen presence in the primitive atmosphere; (2) the lack of geological evidence for the primordial soup of organic compounds or protocells; (3) the high degree of investigator interference in prebiotic simulation experiments; and (4) the very great difficulties in imagining the jump from biopolymers and protocells to the first living and reproducing cell.

Thaxton, Bradley, and Olsen claim an important distinction between *operation science*, which deals with recurring and testable phenomena in nature, and *origin science*, which attempts to understand unique events of the past such as the origin of life or geological upheavals. It is claimed that origin science is much less dependable in generating explanations, and so we should not rule out divine intervention to explain past events. This distinction does not do justice to the *observational-comparative method*, which is heavily used in geology, meteorology, astronomy, and many branches of biology, and is often capable of generating reliable explanations of past events. However, the great time gap and the experimental difficulties involved in origins research place low reliability on attempts at reconstructing the spontaneous origin of life.

The four theistic models for relating science and Scripture are brought to bear on the origin-of-life research. It is concluded that although the desire to find a naturalistic explanation for the origin of life is a major driving force behind most of the research in that area, such an explanation is not capable of excluding God. Whether God employed what we call "natural forces" or "supernatural" processes, life did not originate by chance—God did it.

*Chapter 7*

# THE DARWINIAN REVOLUTION

The Darwinian revolution has been called, for good reasons, the greatest of all scientific revolutions. It represented not merely the replacement of one scientific theory ("immutable species") by a new one, but it demanded a complete rethinking of man's concept of the world and of himself; more specifically, it demanded the rejection of some of the most widely held and most cherished beliefs of western man. In contrast to the revolutions in the physical sciences (Copernicus, Newton, Einstein, Heisenberg), the Darwinian revolution raised profound questions concerning man's ethics and deepest beliefs. Darwin's new paradigm, in its totality, represented a most revolutionary new Weltanschauung [worldview].

ERNST MAYR

## Revolutionary Impact of Darwin's Work

Biological evolution is probably the most controversial and—in some circles—unpopular scientific theory ever advanced. It is continually under attack as being responsible for undermining religious belief and for promoting such evils as communism and sexual freedom. On the other hand, to challenge evolution from within the halls of academia is to risk serious questions about one's sanity.

Why are there such strong feelings about a biological theory? The quote from Mayr suggests an answer: Evolution is more than a scientific theory; it is the starting point for a worldview. This is a worldview that replaces religious belief, and does it in the name of science. The worldview extensions of evolution have been thoroughly entangled with the scientific components, and most evolutionists seem content to keep things this way. But our purposes in this text will be best served if we can untangle and deal separately with scientific and philosophical components of evolution.

### *Elements of the Darwinian Paradigm*

Mayr points out that there are actually five components to the scientific aspects of evolutionary theory as presented by Darwin:

#### *1. Evolution as a Reality*

Evolution as a reality is the view that change has occurred over time in the living world; life is a dynamic feature of the natural world, not the static system pictured by natural theology up to Darwin's time. Of course, this element of evolution was not new with Darwin; Lamarck and others had already challenged the static view of nature. This component of evolutionary theory has given rise to the claim that "evolution is a fact," a claim that is usually countered by opponents who ask (legitimately) how a theory could become a fact.

Commenting on this, Mayr asserts: "For many biologists of today, evolution is no longer a theory but simply a fact, documented by the changes in the gene pools of species from generation to generation and by the changes in the fossil biota in accurately dated geological strata. Current resistance is limited entirely to opponents with religious commitments." In other words, Mayr thinks that the evidence for evolution is so overwhelming that it might as well be considered to be fact.

#### *2. Evolution by Common Descent*

Not only do species change over time, Darwin maintained, they are related to other species by descent from common ancestors. For example, an ancestral weasel could have given rise to the shorttail, longtail and least weasels. But minks, fishers, and martens are in the same family as the weasels (*Mustelidae*), and it is logical to think of all members of this family deriving from a common weasel-type ancestor further back in time. Cats, dogs, weasels, and bears also have many characteristics in common, as carnivores, and they also could have arisen—much further back—from a common ancestor.

This process, thoroughly applied, theoretically links together all organisms back in time to the first living cell. This component of evolutionary theory provided a logical program for systematics (taxonomy) and a rationale for much of the accumulated knowledge in embryology, comparative anatomy, biogeography, and other budding branches of biology.

### 3. *Evolution as a Gradual Process*

The changes that occur in species over time, according to Darwin, are the results of the action of natural selection on very small inherited differences. Even though existing species show sharply distinct and separate characteristics, these are thought simply to reflect a historical process of divergence that has eliminated intermediate forms or connecting species. Over time, enough small changes are believed to occur to produce forms that are quite different from their ancestor species or from other contemporary species that have also evolved from the same ancestor.

### 4. *Natural Selection*

The codiscovery of natural selection by Charles Darwin and Alfred Russel Wallace is the stuff of every biology text. In one stroke, Darwin and Wallace had an explanation for the adaptations of organisms to each other and to their environment, as well as a mechanism that could produce gradual changes in species over time. Although this component of Darwin's theory was slow in gaining acceptance by the scientific community, it now has preeminent status as a creative force in evolution—the arbiter of survival and change in species characteristics. The logic of natural selection involves three inferences, which are based on five "facts" or observations, as follows:

Fact 1. All species have the biological potential to increase their numbers to large populations.

Fact 2. However, populations in nature demonstrate a remarkable constancy in numbers over time.

Fact 3. The resources necessary for the success of a species are limited and are also in relatively constant supply over time.

Inference 1. Competition occurs between members of the same species for the limited resources needed for survival and reproduction. Only a small fraction of the individuals survives and gives rise to the next generation.

Fact 4. No two individuals of a species are identical in the characteristics they show. In fact, the members of a species demonstrate great variability. (Darwin, like all of his contemporaries, was unaware of the mechanisms that govern variability and inheritance of characteristics.)

Fact 5. Much of this variability is genetically based and therefore inheritable.

Inference 2. The competition ("struggle for existence") between members of a species is not random. Survival and parenthood of the next generation depends on the unique hereditary characteristics of the survivors that convey to them an advantage in acquiring the resources for survival and reproduction. This unequal survival is natural selection; the less advantaged are "selected" out and their genes perish with them.

Inference 3. Over time, the accumulation of more favorable characteristics due to natural selection leads to gradual changes in the species, and this is evolution. It eventually leads to the production of new species (Darwin was not completely clear on how this could occur, however).

5. *The Origin of Species*

Darwin was convinced that natural selection was involved in the multiplication of species (usually called *speciation*), and he recognized that geographical isolation of populations was important to speciation—this became apparent to him especially in his studies of the fauna of the Galapagos archipelago. But he did not offer a clear explanation of how one species might give rise to two or more species, in spite of the fact that his earthshaking book bore the title of *The Origin of Species*. This important problem of evolutionary theory was not resolved until the time of the "NeoDarwinian synthesis" in the 1930s and 1940s.

### *The Importance of Darwinism to Biology*

Evolution, as it was introduced in detail by Darwin and as it has been supplemented and modified over time, is the organizing principle of biology. The millions of species of plants, animals, and microbes are classified according to evolutionary principles (common descent). The adaptations of all living things are understood as the products of natural selection acting on genetic variations. The geographical distribution of organisms is interpreted as the outcome of the mechanisms of speciation followed by dispersal. Evolutionary relationships dominate our understanding of the genetics of living organisms, from DNA structure to the appearance of characteristics in the next generation. Ecology, animal behavior, comparative anatomy, embryology—all are organized and interpreted from an evolutionary framework. It is little wonder that Darwin is considered to be the most important person in the history of biology; his work stimulated an explosive growth of research and biological knowledge, and his theoretical contributions stand as a major landmark in the development of human thought.

Consequently, *evolution is a thoroughly biological subject*.

Understanding evolution requires the acquisition of a full-fledged biological education as well as some intensive study in evolutionary thinking. Unfortunately, most of the critics of evolution have only a shallow or narrow knowledge of biology, and their writings on the subject of evolution reflect it. These are often physical scientists, engineers, philosophers, or theologians who are reacting to the philosophical extensions of evolution. Convinced that biological evolution must be false because of some of the claims of philosophical evolutionism, many critics direct their attention to the biological arena, where evolution is so thoroughly embedded that attacking it is like attacking most of biological thought.

Until quite recently, these attacks on evolution were ignored by all but a few Christian biologists who were interested in separating the philosophical and theological issues from the scientific ones. However, in the last few years, opponents of evolution have had remarkable success by promoting the concept of equal time for creationist and evolutionist models of origins in the public schools. Several book-length challenges to evolution have taken the tactic of critiquing evolutionary ideology at its weak points, while not appearing to promote a religious alternative. These new tactics have drawn an often angry and well-orchestrated response from defenders of evolution in the mainstream of science. At the very least, all of this attention underscores the fact that Darwin indeed touched off a revolution.

### *The Darwinian Revolution and Human Thought*

Clearly, biology could never be the same after Darwin's work. Chapter 4 described how evolution by natural selection represented a new paradigm for major areas of biology, one that no longer rested on biblical assumptions. In place of design by God, natural selection explained how adaptations could arise and be so finely tuned to the environment: A moth could strongly resemble the tree trunk it rested on because natural selection favored those individuals which over time were protected by their coloration, not because God designed it that way in the original creation. Once design was removed as a secondary (immediate) cause for the observed data, purpose was no longer a logical first cause. If God didn't design it, then the purpose of an adaptation was no longer seen as an example of God's wisdom and loving care for his creatures. Thus teleology—the assignment of purpose to elements of the cosmos as an explanation for their origin—was removed from biology.

So the impact of the Darwinian revolution reached far beyond biology. As this chapter's opening quote shows, evolution was and continues to be seen as a serious challenge to religious faith. In the place of design and purpose, Darwinian thought offered natural selection and the autonomy of nature. Instead of creation of living forms by God a few thousand years ago, Darwin offered the origin of species by natural forces over vast reaches of time. Darwin's view of evolution by common descent linked human beings with the other primates and, by extension, with other mammals, suggesting that humankind and the apes shared common ancestors. According to Mayr, "The claim—or, one might say more correctly, the demonstration by science—that man was not a separate creation but part of the mainstream of life caused a tremendous shock." Mayr sums things up: "By providing a purely materialistic explanation for all phenomena of living nature, it was said it (natural selection) 'dethroned God.'"

These, of course, are worldview claims. And they are thoroughly embedded within a framework that appears to pit science against religious belief (guess who's supposed to win?). As we have seen in earlier chapters, this posing of events and processes in nature as either/or phenomena does justice to neither science nor biblical teaching. It is quite apparent from

Mayr's analysis that he considers the worldview extensions of Darwin's work as *the* Darwinian revolution, and here Mayr shows his naturalistic bias. On the contrary, I would suggest that the impact of Darwin's work on the biological sciences was his greatest and most revolutionary accomplishment. Although he was fully aware that evolutionary thought challenged the static view of nature derived from Genesis, Darwin took no delight in the acrimonious debates that developed. He was victimized by what philosopher Karl Popper later called "the law of unintended consequence," as his scientific work began to influence other areas of human thought far afield from biology. Darwin's own views are interesting, as this quote from a letter written in 1879 shows:

> It seems to me absurd to doubt that a man may be an ardent theist and an evolutionist. . . . What my own views may be is a question of no consequence to anyone except myself. But, as you ask, I may state that my judgment often fluctuates. . . . In my most extreme fluctuations I have never been an atheist in the sense of denying the existence of a God. I think that generally (and more and more as I grow older) but not always, that an agnostic would be the most correct description of my state of mind.

Since much of this book is designed to deal with the philosophical and religious issues raised by the Darwinian revolution, we should not attempt to do it all right here. Instead, let's turn our attention to the ways in which Darwin's work is interpreted today from within biology.

### Evolutionary Thought Today

Biological texts generally seem to organize their coverage of evolution into four categories: (1) *evidence for evolution*—that is, the evidence that evolution has occurred; (2) *phylogenetic questions and ideas*—the realm of macroevolution, or evolution above the species level; (3) *the mecha-*

*nisms of evolution*—the processes that have produced adaptations, and the array of species; and (4) *the historical development of evolutionary theory*—beginning with Darwin's work. Of these four categories, we have already given enough attention to the last. Let's look briefly at the first three.

### Evidences for Evolution

The direct evidence for evolutionary change consists of two elements: (1) *the geographical relationships of species*, and (2) *the fossil record*. Both Darwin and Wallace were strongly influenced by their studies of the spatial distribution of plants and animals. Wallace observed: "The most closely allied species are found in the same locality or in closely adjoining localities and . . . therefore the natural sequence of the species by affinity is also geographical." *Speciation*, the process of splitting a parent species into two or more daughter species, is a well-established process that is often called *microevolution* because it ordinarily involves minor differences between the species in question. However, this process may be read backwards to yield the concept of *common descent*, and very quickly becomes a strong argument for the evolutionary linkage of taxonomic units above the species level and the construction of phylogenetic trees.

The fossil record, on the other hand, yields evidence for *vertical evolution* (evolution over time) by revealing the prior existence of forms now extinct but similar in many ways to existing species. In addition, general trends in the fossil record suggest a progression: For example, simple cells are the only forms found in the oldest fossil-bearing layers; invertebrates are found in the Cambrian (early Paleozoic), the first reptiles in the late Paleozoic, the first mammals in the Triassic, and placental mammals in the late Cretaceous. The very old age of the earth and the long history of life forms also support evolution, in that these are necessary to the theory.

Indirect evidence for evolution comes from a number of

biological subdisciplines. *Morphology*—the study of animal and plant form—documents the similarities and differences between species. Common descent assumes that present-day forms have inherited similar structural patterns from common ancestors, and thus provides a basic program for morphological studies. Those studies provide confirmatory evidence for evolution, in that evolution is presently the only available scientific explanation that can make sense of the evidence from comparative anatomy. *Embryology*—developmental patterns in animal forms in particular—indicates that certain developmental pathways suggest ancestral relationships. For example, there is the appearance of gill arches in land animal embryos, and a notochord in higher vertebrates.

*Genetic studies* have demonstrated the significant plasticity of species, a necessary property if species are subject to natural selection and can diverge into daughter species. This is a consequence of variation in the genes that determine a given characteristic, and the great variety of different expressions of the collection of genes that make up an organism. *Molecular biology* has uncovered the genetic code, which is essentially identical in all organisms and therefore points to the possibility of tracing all of life on earth back to a single origin. These are merely examples of a large body of biological knowledge that can be tied together (not without some knots, as we will see) by the use of evolutionary explanations—hence it supports evolution. As we have indicated, modern life science reflects the evolutionary paradigm in foundation and detail so thoroughly that it is difficult to imagine a biology founded on anything else.

### Microevolution to Macroevolution

*Microevolution* ordinarily refers to the minor changes that accompany speciation processes and short-term adaptations to changing environments. They are the outcome of genetic variation—from a variety of sources—within a species or population, acted on over time by natural selection. The changes involved might be differences in color or pattern, or small size differences, or minor changes in behavior. They are sufficient to render a population or species distinctively different from another, closely related population (speciation). Or, the changes might represent a favorable adaptation to the environment and so bring about a change in the genetic makeup of a single species over time without any implications for speciation.

*Macroevolution* refers to evolution "above the species level," a somewhat confusing concept, since all evolutionary change is believed to occur as a consequence of microevolutionary processes acting on species. In effect, macroevolution looks back along the lines of common descent, using all of the resources of the observational-comparative method, in an effort to reconstruct the pathways of descent and therefore the evolutionary relationships of organisms (in other words, their *phylogeny*). Comparative anatomy, embryology, and paleontology are the major data sources for this approach, although molecular biology has begun to provide some interesting data in the form of protein structure and the fundamental sequences of the genetic code.

Macroevolution is highly retrospective, therefore; it employs whatever fossil evidence is available, and requires the extensive use of inference in the development of explanations. The data are sketchy, the "transitional forms" linking different taxa are often lacking, and so the entire enterprise is developed within a cloak of uncertainty. Many problems that were apparent to biologists of the 1800s still exist: How are the different orders of birds related? Which of the protozoan groups gave rise to the metazoans? How are the different invertebrate phyla related?

Microevolution and macroevolution are part of the fabric of evolutionary theory; they are based on the same mechanisms, and one is clearly an extension of the other further into the past. Although they rest on different data bases, and one (microevolution) can be assigned a higher probability than the other, the explanation of phylogenetic relationships between organisms represents a continuum in reasoning that is hard to break. Let's look at an example.

We have little trouble accepting the (micro)evolutionary relationships of members of the genus *Dendroica* among the New World wood warblers placed in the family Parulidae and the order Passeriformes. Some twenty-two species are placed in this genus, and the similarities between the species are obvious. What about the genus *Oporornis*, also in the family Parulidae? There are four species in this genus, and they are slightly larger than the *Dendroica*, have shorter tails, and tend to exploit habitats close to the ground—different, but not by much. Is this macroevolution or microevolution? The family Paridae (chickadees and titmice) is also in the order Passeriformes, along with twenty-six other families. These are all "perching birds," with many characteristics in common. Common ancestry is easy to picture for the members of the same genus, not much more difficult for the genera placed within a family, and not too difficult to imagine for the different families within an order. Similar sequences of possible relationships can be found in practically every group of plants or animals; this is one reason why taxonomists and ecologists are often among biologists the most supportive of evolutionary theory.

Clearly, it is much more difficult to reconstruct connections between major taxonomic categories like phyla or classes. There are major discontinuities in the existing "tree of life," and in the fossil record. The separations occurred in the distant past. Plausible explanations may be given for these discontinuities, but their existence makes reconstruction of evolutionary history an exercise in probability.

It is quite common for critics of evolution to "accept" microevolution and "reject" macroevolution, believing that it is only macroevolution that is in conflict with the Genesis account of origins. The greater uncertainties in the development of macroevolutionary explanations makes this distinction appealing to the casual student of evolution. But it is a distinction that is made for religious reasons rather than scientific ones. As we have seen, breaking into the continuum of phylogeny is an arbitrary act. And if macroevolution is challenged, what is there to erect in its place as a paradigm for explaining the data that have accumulated over the years in so many subdisciplines? Before we look more deeply into the challenges to evolution, we should take notice of the current level of agreement about evolutionary theory reached after many years of controversy—the Neo-Darwinian synthesis.

### *Mechanisms of Evolution: The Neo-Darwinian Synthesis*

In the years following publication of *The Origin of Species*, evolutionary biologists gradually became aligned into two opposing camps. The naturalists (field biologists, taxonomists, and ecologists) placed a strong emphasis on gradualness (the result of natural selection) as the mode of evolutionary change. They frequently espoused a form of Lamarckian inheritance (inheritance of acquired characteristics), as had Darwin in some of his writings. The geneticists, following the discovery of Gregor Mendel's crucial work, emphasized mutation pressure as the primary driving force of evolutionary change and minimized the roles of natural selection and recombination.

By the late 1940s, the controversy was resolved as both camps recognized the implications of geography and populational genetics for evolutionary change. Species were recognized as populations in space and time, configurations of highly organized but still variable genetic material (the gene pool). Mutation was acknowledged as the ultimate raw material on which natural selection could operate, but the Mendelian recombination of genes in reproduction was recognized as the immediate source of genetic variation. Evolution was therefore seen as a gradual process of small genetic changes brought about by mutation and recombination acted on by natural selection. Such gradual changes were accepted as sufficient explanation for macroevolutionary trends, given enough time. Speciation was understood as the consequence of geographical and reproductive isolation in a context of ecological factors that also varied and constituted the natural selection that operated on the gene pool. Current biology texts basically reflect the consensus brought about by this reconciliation.

However, this consensus does not go unchallenged. Evolutionary theory is subjected to continual examination and revision by its supporters, and to criticism and challenge by its detractors.

## Scientific Challenges to Darwinian Evolution

It is not my purpose here to participate deeply in the debates about creation and evolution. There is already an immense literature covering this subject from many viewpoints. I would like to focus on a few problems that are raised by critics and supporters alike, however, in order to present evolutionary theory as it exists today.

### *The Fossil Record*

Fossils are the only truly historical evidence for evolution, yet the fossil record is a record of discontinuity rather than the continuous change suggested by Darwinian gradualism. Gaps are more common than transitional forms; mass extinctions occur; new, complex forms appear quite suddenly in the fossil record; other forms exist for millions of years without appreciable change. Several explanations are offered for this state of affairs:

1. The fossil record is a haphazard collection of forms that happened to get trapped in sediments—it is a very imperfect record of the history of life. As evidence of this, several entire phyla of animals are known only from one location and by one or several species.
2. The gradual changes postulated by Darwin are not the major mode of speciation; instead, new species developed rapidly as subpopulations became geographically isolated from the mainstream species population, reaching and maintaining equilibria for a more extended time until conditions promoted another burst of speciation—the *punctuated equilibrium* theory of Niles Eldridge and Stephen Jay Gould. In this view, the transitional forms appear as an instant in geological time, and only the long-lived, dominant species are significantly recorded in the fossil record. Thus the fossil record is a much better record of evolutionary change than previously thought. This theory is a substantial departure from the gradualism of Darwin, and is currently the center of a major disagreement between evolutionists about the dominant mode of evolutionary change.
3. Special creation by God is another explanation offered by some Christians to account for the fossil record. In this case, the gaps in the record are presumed to be signs of genuine breaks in the continuity of life forms. This explanation involves "bursts" of direct creative acts by God at crucial periods in the history of life (progressive creation). New forms are essentially the Genesis "kinds," and these forms diverge over time along the lines of microevolutionary theory.

One of the most extraordinary examples of discontinuity is the so-called *Cambrian explosion*, the appearance of representatives of all major marine invertebrate phyla in the earliest sedimentary fossil-bearing rocks (dated at 570 million years before present) as well as a variety of other phyla that have not left descendents. The only possibly ancestral forms that have been found are from the *Ediacara complex* (dated around 700 million before present), some soft-bodied animals resembling the annelids and cnidarians and some creatures that seem unconnected to anything appearing later in the fossil record. Explanations offered for the lack of fossils prior to the Cambrian are to date not very satisfactory (for example, there was an explosion of speciation as ecological niches were created and filled; all of the ancestral fossils were soft-bodied and were not preserved). We simply do not know where or how these well-developed invertebrates originated, and this is often a part of the story that is omitted from the texts.

However we interpret the Cambrian explosion, it is only one of many discontinuities in the geological strata. Mass extinctions occurred at the end of the Permian, for example, when over half of the existing families of marine organisms disappeared. Again and again, there are indications of mass extinctions (where did the dinosaurs go?) and the appearance of new fauna and flora. The present array of plants and animals bears little resemblance to the earliest forms; new groups, such as reptiles and mammals, appear and increase in dominance over time as numbers of species increase. The serious questioner must deal with these realities as well as the gaps and the explosions.

What explanation can best account for all of the observed data? It seems clear that the fossil record, taken alone, is insufficient as a data base on which to build a coherent explanation for the history of life. It presents problems for evolution, and for the creationist alternatives. In the case of evolution, the problems stem mainly from a lack of fossil evidence—the absence of transitional forms. For "scientific creationist" interpretations, major problems are posed by the existing fossil record, which gives overwhelming evidence of great age and many changes.

### *Other Criticisms*

One interesting challenge to evolution comes from mathematics. It has been claimed that a mathematical analysis shows the impossibility of producing all known complex living systems by the accidental shifting and rearrangement of mutations, even granting the nearly 5-billion-year age of the earth. Summarizing the findings of a conference on the mathematical probability of the evolutionary theory held in 1965, a group of mathematicians reported: "It is our contention that if 'random' is given a serious and crucial interpretation from a probabilistic point of view, the randomness postulate is highly implausible and that an adequate scientific theory of evolution must await the discovery and elucidation of new natural laws."

A subsequent analysis of this charge by a group of evolutionists pointed out that the mathematicians greatly oversimplified the biological processes involved. Their view of genetic variability did not include a proper recognition of the many ways new genes and groups of genes may be produced in organisms. Indeed, studies of the immune system of mammals have revealed the amazing capacity of genetic systems to rearrange genes in a way that leads to new properties. The evidence for this is strong; many enzymes with quite different functions share identical gene pieces. The mathematical problem is fundamentally changed if, as is the case, new combinations occur as a result of the shifting around of pieces of genes rather than single bases. Further, large segments of DNA can be shifted between chromosomes and even between different species—the *transposons*. The mathematicians also failed to take account of the fact that genetic changes take place simultaneously at hundreds or thousands of gene locations in an organism.

By placing its emphasis on the role of chance or randomness, the mathematical approach also underrates the highly nonrandom role played by natural selection. In short, it appears that the purely mathematical approach to evolutionary theory is far too simplistic to provide a plausible critique of the theory.

Another challenge to evolution attacks the validity of the theory as science. There are several elements to this challenge, of which we will examine two. One is the observation that Darwinian evolution is not a scientific theory because it is not capable of being tested experimentally; it is retrospective in principle. This is the equivalent of the distinction between operation science and origin science from chapter 6, and we saw in that chapter that the falsifiability principle is not an appropriate guide to what is truly science. Furthermore, as we have seen, evolution is a complex paradigm, consisting of mechanisms that operate in the present as well as reconstructive explanations of the past. The mechanisms are indeed testable by experiments, both in the laboratory and in the natural world. Gene mutation and recombination are thoroughly researched and understood; selection has been shown clearly to be capable of bringing about remarkable changes in the genetic makeup of a group of organisms (both artificial and natural forms of selection). And origin science events are subject to the observational-comparative method, which is certainly a valid approach for science.

Some critics also say that evolution is not a legitimate science because its major mechanism, natural selection, is a tautology (its definition involves circular reasoning). Natural selection claims that some variants are better fitted to the environment and will survive and reproduce because of their "fitness." But how do we define fitness? Simply as those individuals that survive and reproduce. Evolutionists answer this criticism by pointing out that fitness is a highly definable property that can be understood as a consequence of differences in reproductive ability, differences in the characteristics that organisms possess, and differences in the environment that predict which characteristics are adaptive; these can be matched to the characteristics shown by existing and surviving members of a species.

These are some of the most significant challenges to evolution from scientists with no "creationist" axe to grind. These criticisms from within the conventional scientific community are signs that evolutionary theory is not a monolithic dogma that goes unchallenged; quite the contrary. Yet it must be said that even those who criticize the theory are rarely attempting to pull it down; their objective is usually corrective.

On the other hand, there are critics who firmly believe that not only is evolution wrong, it is also atheistic and intentionally destructive to belief in God—it is inconsistent with biblical Christianity. Let us turn our attention to this subject.

## Christian Responses to Evolution

### *Evolutionism*

Up to now, we have been dealing primarily with evolution as a biological theory. It is possible, however, to adopt a worldview in which evolution plays an important role, and then evolution becomes *evolutionism*—the belief that a natural process, evolution, has been entirely responsible for the development of all of life as we now know it. There is no purpose behind this development, no direction or design to be perceived in the natural world. The evolutionary process is therefore seen as a substitute for divine action—in effect, replacing God. It becomes a philosophy, or even a religion, as it is used to interpret history, sociology, ethics, and religion itself. Evolutionism is thus a form of naturalism, whose basic belief is that ultimate reality consists entirely of matter and its space-time arena. Again we encounter Nature (with its capital N) as autonomous—a self-originating and self-sustaining reality that has no purpose but accomplishes everything; as popularly presented, Mother Nature and her marvelous accomplishments.

One Christian response to this sort of thinking is to claim those accomplishments and indeed the processes themselves for God—and that is entirely appropriate. As Christians, we should insist that all processes occurring in the natural world are governed by God as an outcome of Creational Law. We need on the one hand to deny the worldview extensions of evolution, and on the other to claim the evolutionary processes as part of God's activity in his world. There are many Christians who are convinced that Darwinian evolution is God's basic creative method, and that there is purpose and design in the living world because life has developed exactly as God intended. However, there are many Christians who do not agree with this view, and we need to see why.

### *Nature of the Creation/Evolution Controversy*

Having now looked thoroughly at Genesis 1 (in chapter 5) and at the testimony of the rest of Scripture concerning Creation and the identity of the Creator (in chapter 2), we are better equipped for putting this controversy in perspective. Even though "creation," for mainly historical reasons, is associated with miraculous and instantaneous events, biblical theism views creation as an eternal covenantal relationship between God and his creation. We saw in chapters 2 and 5 that the Bible does not indulge in explanations of how God created and how he sustains his creation. Scripture clearly answers the questions of who created the cosmos, and why—not how, not even when. Therefore *the real issue is Creator versus no Creator, not creation versus evolution*. Is the God of the Bible real? If so, then he creates all things—in his way and his time, and for his purposes. The alternative is to believe that all things are the product of purposeless mechanisms that somehow arose spontaneously as a consequence of the appearance of matter and energy (and where did these come from?).

The concepts of Creation and evolution are answers to questions at two different levels; They are not mutually exclusive alternatives, as Table 3 indicates. At the level of worldview beliefs, we have no Creator (a naturalistic philosophy) or Creator (theistic); at the level of scientific explanation, we have evolution by common descent or instantaneous origin. Questions appropriate to the upper level (who, why) are a matter of faith and belief—worldview considerations. Questions appropriate to the lower level (what, when, how) are a matter of looking carefully at the evidence and deciding which model best answers the questions. As Fischer has pointed out, acceptance of an explanation at the worldview level does not necessitate acceptance of the alternative appearing below it. As an illustration, recall that one explanation offered for the origin of life (in chapter 6) was panspermia, or the introduction of the first life form from outer space. This is an example of an instantaneous origin that is compatible with a naturalistic world view.

*TABLE 3*

*Format for Resolving Creation/Evolution Problems*

| Questions: | No creator (naturalistic) | or | Creator (theistic) | Level of worldview |
|---|---|---|---|---|
| who, why | | | | |
| what, when, how | Evolution by common descent | or | Instantaneous origins | Level of science |

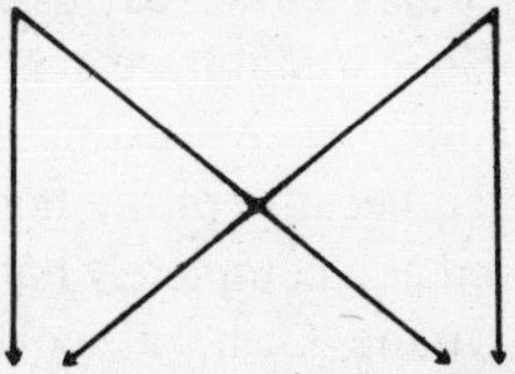

A theistic worldview does not require any particular explanation of origins. The choice is a matter for scientific reasoning and sifting of evidence. Remove evolution from the naturalistic worldview assumptions that are usually tied to it and it no longer poses a serious threat to biblical faith. Remove the scientific creationist or instantaneous origins view from a worldview status and it faces examination as a model at the scientific level. What happens then?

### *Problems with the "Scientific Creationist" Model*

We have previously labeled the scientific creationist model as substitutionism, which proposes that there is a scientific explanation for origins that is taught by Scripture, and therefore this explanation—referred to as *creation science*—is a legitimate substitute for the interpretations of origins coming from the scientific community (thus evolutionary). What, then, is creation science?

Creation science states that the earth and living things were created in six twenty-four-hour days some 6,000 to 10,000 years ago. Since the creative acts were one-time miraculous events, they cannot be considered within the framework of natural mechanisms as we know them. Creation science considers the only real evidence of origins to be the fossil record. Since this record contains many gaps, it supports the view that God created living forms (the Genesis "kinds") as they appear to us today either as fossils or as present-day species. The geological strata are essentially a record of the Noahic flood, and the massive extinctions recorded in the fossil record occurred because of the flood.

Evolution is viewed as a completely naturalistic mechanism; in the words of Duane Gish, a spokesman for Creation science:

> The theory of evolution is an attempt to provide an entirely mechanistic, naturalistic explanation for the origin of the universe and all

that it contains. . . . The claim that God used evolution rather than special creation to bring the universe and the living things it contains into being denies the omniscience and omnipotence of God and makes a mockery of Scripture.

Their writings indicate that creation science mainly involves amassing criticism against most aspects of evolutionary theory. The major source of the criticism is the writings of other evolutionists who are challenging various elements of the theory in order to correct it. Along with flood geology, creation science holds to a "young earth," and therefore criticizes the efforts of mainstream science to establish an ancient age of fossils and of the earth and universe.

The problems with Creation science are many and serious. Several stand out:

1. It denies the massive evidence for an ancient age of the earth and universe. Christian geologist Davis Young comments:

   Since most creationists [meaning scientific creationists] think Scripture teaches that the earth is young, they enthusiastically believe and advance such arguments. However, it has been shown that these scientific arguments for a young earth are not valid, and they do not establish a young age for the earth at all. These so-called scientific evidences are based on incomplete information, wishful thinking, ignorance of real geological situations, selective use of data to support the favored hypothesis, and faulty reasoning. The fact of the matter is that the scientific evidence considered as a whole, and as we have it now, compellingly argues for the great antiquity of the earth.

2. Creation science inserts a dualistic wedge between God and natural processes by stating that evolutionary processes are mechanistic and therefore naturalistic and not God's doing. This is more serious than it might look; the implication is that unless a process is miraculous, supernatural, it could not be of God—a view that puts us

back into the nineteenth century and the days of natural theology. A corollary to this view is that God creates only by supernatural means; living things thus first appeared on earth fully developed and much as we know them today. As we have already seen, this is a woefully inadequate view of the meaning of creation.

3. Creation scientists do not listen to criticism, even when it comes from fellow Christians. Again, Davis Young states the case:

> I find it puzzling that the spurious arguments of creationism are repeated over and over again in so many books, articles, and periodicals. Creationists need to learn how to receive criticism when they are told that they have spoken on a matter about which they know relatively little. . . . The creationist movement seems unwilling to receive criticism in a gracious manner and profit from criticisms of its arguments by competent Christian scientists. The same arguments are put before the Christian public in book after book, in article after article. Christian people are still being misled.

4. Creation science insists that biblical Christianity requires rejection of evolution and acceptance of its views. When it does this, the creation science movement is adding to the gospel a belief that consists of uninspired interpretation. This can become a serious stumbling block in the way of unbelievers, particularly when their view of Christianity comes primarily from their contact with the creation/evolution controversy.

Creation science generates intense reactions from the scientific community. Here is the comment of an evolutionist toward the end of a book defending evolution against creationism:

> I believe Creationism is wrong: totally, utterly, and absolutely wrong. I would go further. There are degrees of being wrong. The Creationists are at the bottom of the scale. They pull every trick in the

book to justify their position. Indeed, at times, they verge right over into the downright dishonest. Scientific Creationism is not just wrong: it is ludicrously implausible. It is a grotesque parody of human thought, and a downright misuse of human intelligence. In short, to the Believer, it is an insult to God.

It is sad to have to read this commentary on a subject that ought to be leading people toward faith rather than away from it.

### *Final Thoughts*

Evolution as an explanation of origins is not a fact, and is not even a theory. It is a biological concept, and a very complex one. It is a major paradigm for biological thought and research, based on data and logical reasoning, yet requiring a great deal of creative guesswork that may never be verified. It has weaknesses: Uncertainties are always involved in extrapolation to the past; the fossil record is incomplete; there are loose ends, especially in view of the information coming from molecular studies of gene structure and control; proponents constantly overstate their case, often confusing a scientific explanation with a worldview. But there is no scientifically useful alternative to evolution; and if it is to be evaluated, the evaluation should take place in the scientific arena. No one should accept evolutionary reasoning without examining the evidence, but be warned: The evidence is strong, and it is convincing.

Was Darwin basically correct in his analysis of evolutionary change over time? Many biologists who are Christians are convinced that he was, and therefore would be comfortable with the view that evolution is God's creative method. In this, too, they would be in agreement with Darwin, as the last paragraphs of *The Origin of Species* indicate:

> To my mind it accords better with what we know of the laws impressed on matter by the Creator, that the production and

extinction of the past and present inhabitants of the world should have been due to secondary causes, like those determining the birth and death of the individual. . . . There is grandeur in this view of life . . . having been originally breathed by the Creator into a few forms or into one; and that . . . from so simple a beginning endless forms most beautiful and most wonderful have been, and are being evolved.

## Summary

Biological evolution is a highly important scientific paradigm that is often employed in nontheistic worldviews. The five components of evolutionary theory are described: evolution as a reality, evolution by common descent, evolution as a gradual process, natural selection, and the origin of species. The evolution paradigm is the primary organizing principle of biology; understanding evolution requires a thorough biological education. Recent attacks on the theory (usually by nonbiologists) have brought an angry response from its defenders.

The Darwinian revolution has been viewed as a threat to religious faith because it appeared to replace design and purpose in the natural world with natural selection and an autonomous nature. In place of creation of living forms (especially human beings) by God, evolution presented the origin of species by natural forces. Although some writers believe that Darwin's major contribution was in the worldview arena, it can be argued that his contribution to biological theory was the true Darwinian revolution. Darwin himself was sympathetic to theistic views, but remained an agnostic.

Current presentations of evolutionary thought include (1) A thorough description of the evidences for evolution—including the geographical distribution of species and the fossil record (direct evidence), and indirect evidence from many of the biological disciplines. (2) Phylogenetic questions and ideas—in particular, *macroevolution*. This is seen as forming a continuum with *microevolution*, so that the relationships between forms can be extended into the past along lines of common descent. Although many try to separate these forms of evolution for religious reasons, it is arbitrary to attempt to break into the theory in this way. (3) Mechanisms of evolution—the Neo-Darwinian synthesis represents a consensus that is usually presented by current texts.

Some problem areas of evolutionary thought are discussed. The fossil record, with its many gaps, can be viewed as very incomplete or as evidence for a mode of evolutionary change called *punctuated equilibrium*, in which transitional forms are very short-lived (in terms of geological time) and only the long-lived equilibrium forms are ordinarily found. A third interpretation is that of special creation, suggesting that the gaps in the record are real discontinuities between life forms. Other criticisms of evolutionary theory are discussed: mathematical improbability, the "nonscience" of the study of past events, and the redundancy in the idea of natural selection. These do not appear to be serious problems to evolutionists.

The problem of evolutionism is raised as a legitimate Christian concern. Here evolutionary theory supports a worldview that sees nature as autonomous. The biblical view holds that God governs the natural world through Creational Law; it is not autonomous. Thus evolution may be God's creative method, and life has developed exactly as God has intended. The real nature of the creation/evolution controversy is therefore the issue of Creator versus no Creator, and these are clearly belief or worldview concerns that speak to the questions of who or why. Questions of what, when, and how are questions at the level of science, to be answered by scientific examinations of the evidence.

*Creation science* is summarized and evaluated from the perspective of several of the major problems it presents: (1) It denies the massive evidence for the age of the earth and universe. (2) It forces a wedge between God and his world by suggesting that anything that operates "mechanistically" in the natural world is not of God. (3) The creation scientists do not appear to benefit from well-reasoned criticism. (4) Creation science ties itself to Christian belief in general and thus represents a stumbling block to scientists outside of the faith.

Concluding remarks suggest that Darwin was basically correct in his analysis of evolution, and that the evidence for evolution is strong and convincing. It should not, however, pose a problem for theistic belief—for if it happened, God was responsible for it.

*Chapter 8*

# WHERE ARE YOU, ADAM?

A . . . cry of outraged indignation greeted Darwin's *The Descent of Man*. The point was not simply that human dignity seemed to be impugned by Darwin's theory of man's origin, although that was injury enough. Worse yet, the whole doctrine of inspiration, with its comforting assurance that everything man needed to know about his origin, duty, and destiny had been divulged by God himself, was shaken to its very foundations. The great theological doctrines of creation and fall, sin and redemption, with all the moral precepts hanging on them, seemed threatened by a preposterous theory that man was a cousin to the apes, if not of closer consanguinity.

JOHN C. GREEN

## The "Death" of Adam

We now cross the threshold of a highly disturbing subject—our own origin. What about us? Did humankind also evolve? This was the key question in the evolutionary debate following the publication of Darwin's *The Origin of Species*. Common descent leads to a phylogenetic tree that relates all organisms according to the available morphological and fossil data. Our place on the tree (check your biology text!) is in the order Primata, superfamily Catarrhini, family Hominidae. Perched quite close to us in the tree (in our superfamily) are the Old World monkeys and baboons, the gibbons, and the great apes (family Pongidae). Aware that this was the most controversial element in evolution, Darwin postponed making a statement about human beings until 1871, when he published *The Descent of Man*. In this book, Darwin fulfilled the suspicions of his contemporaries by asserting that people had evolved from ape-like ancestors.

T. H. Huxley had already taken up the cause, however, in his famous interchange with Bishop Wilberforce in 1860 and the publication of *Man's Place in Nature* in 1863. The debate itself has earned a place in history, not for having settled anything but for the sharp exchange between Wilberforce and Huxley. In this exchange, Wilberforce reportedly tried to ridicule evolution by asking Huxley if Darwinists traced their descent from monkeys on their grandmother's or grandfather's side of the family. Huxley is believed to have replied, with customary wit, that he would prefer to have an ape for an ancestor than someone who, like Wilberforce, used his great gifts of rhetoric to obscure the truth. (Harvard biologist Steven Jay Gould recently examined the historical evidence for this event, and found that although something like the reported exchange took place, Huxley was not debating with Wilberforce and in fact some observers felt that Wilberforce got the best of the exchange.) However, with the publication of his book in 1863, Huxley threw down the gauntlet by pointing to the physical similarities between human beings and the great apes and arguing for our descent from some extinct common ancestor.

*The Death of Adam* is the title John Greene gave to a book tracing evolutionary thought through the 200-year period before Darwin. Greene meant nothing irreverent by this title; he was simply saying that as the sciences developed during this crucial period, popular belief in the historical accuracy of the Bible gradually declined. The Old and New Testaments clearly regard Adam and Eve as the parents of the human race. Instead of Adam, Darwin and Huxley were offering an ape-like ancestor, and a mechanism for "creating" humankind that seemed to leave God out of the picture. The last bulwark of 18th century natural theology—the uniqueness of the human being as created in God's image—began to crumble in the face of a plausible explanation for the development of our unique capabilities as a result of natural selection. As the opening quote from Greene shows, the very doctrine of inspiration of Scripture appeared to be threatened by the idea of human evolution.

Many of Darwin's contemporaries were not satisfied with the evidence for humankind's evolution. The only relevant fossils unearthed by 1871 were Neanderthals, and these were far too close to modern humans to be considered "the missing link" that would clinch the case. Even Alfred Russel Wallace, the codiscoverer of natural selection, found it difficult to imagine how the unique human characteristics—human intelligence, our moral and religious dimensions—could have evolved. Wallace argued for intervention by God in the process of human evolution: "a superior intelligence has guided the development of man in a definite direction, and for a special purpose, just as man guides the development of many animal and vegetable forms."

And so a stalemate ruled over the highly controversial question of human evolution until the end of the nineteenth century. By that time, the authority of Scripture was under attack more from within the Christian church than from evolutionary biology; higher criticism and comparative religion fueled the fires of liberalism, and the conservative elements of Protestantism responded with a famous series of pamphlets called *The Fundamentals*. Fundamentalism was born, but this movement paid little attention to evolution until the 1920s.

## The Evidence for Human Evolution

How do things look now that 100 years have passed and we have experienced the explosive development of biological knowledge? The search for our possible ancestors began in earnest with the discovery of "Java man" by Eugène DuBois in 1891, and today represents one of the most well-publicized fields of scientific endeavor. Lucy, the Olduvai Gorge, Donald Johanson, and the Leakeys are familiar names. Many species of hominids have turned up in the fossil record, and these fossils are now considered as strong evidence for human evolution. To this is added the most recent evidence from the

DNA and proteins of current primates, which bear further witness to the similarities between humans, gorillas, and chimpanzees that so impressed Huxley a hundred years ago. Is there any hope for Adam now? Once again, it is the Christians in biology who are in the eye of the storm. It is our discipline that is responsible for most of this disturbing information, and we are obliged to deal with it. The fossil record is a good place to start.

### *The Fossil Record: Lines, Branches, and Dead Ends*

A reading of biology texts will introduce you to a group of hominids in the genus *Australopithecus*, and another group in the genus *Homo*. The Australopithecines were smaller than modern humans, walked upright, but were more like apes from the neck up. *Homo habilis* and *Homo erectus* were larger animals and had skulls and dentition much more like humans than apes, and were tool users. Most of the oldest forms come from Africa, now considered to be the cradle of hominid evolution. Many *Homo erectus* fossils have been found in Asia, a few in Africa. The possible relationships and dating for these hominids are indicated in Table 4. This is only *one* current interpretation of the fossil record of hominids. (Stephen Gould reports that each year prior to his lecture on hominid fossils he dumps out his file contents and starts over again.) As Table 4 indicates, there are many gaps in the record, and some significant overlaps—the most important being the overlap between *Homo habilis* and the Australopithecines. Thus the genus *Homo* may be as old as *Australopithecus*.

The differences of opinion regarding this group of fossils are interesting. The National Academy of Sciences asserts that the "missing links" have been found; "a succession of well-documented intermediate forms or species" tie early primates directly to humans. Ernst Mayr speaks of an almost unbroken chain from the earliest *Australopithecus* to *Homo sapiens*.

TABLE 4

*Fossil Hominids, Dating and Brain Capacity*

| Age in millions of years | Species and brain capacity (cc) |
|---|---|
| 0.03 | *H. sapiens* (modern) (1,350) |
| | (overlap) |
| 0.05 | *H. sapiens* (Neanderthal) (1,600) |
| | (gap) |
| 0.3 | *H. sapiens* (archaic) (1,200) |
| | (gap) |
| 1.5 | *H. erectus* (1,000) |
| | (overlap) A. *boisei* (530) |
| 2.0 | *H. habilis* (680) (gap) A. *robustus* (530) |
| | (gap) |
| 2.5 | A. *aethiopicus*? (410) |
| | A. *africanus* (440) |
| | ? |
| 3.0 | |
| | (gap) |
| 3.5 | A. *afarensis* (400) |

Those closer to the field are not so positive, however. Some anthropologists consider A. *afarensis* to be ancestral to the *Homo* line as well as later Australopithecines. Others believe that the genus *Homo* developed independently. Gould points to three coexisting lines of hominid fossils (A. *africanus*, robust Australopithecines, and *Homo habilis*), none of which show any trends toward change during their appearance in the record. In fact, Gould states that no progressive changes are found in the fossils of any hominid species. One very significant gap in the record is the lack of any form that might be ancestral to both hominids and the modern apes. Future finds may fill in the gaps; at present, however, the data base is far too limited for constructing "an unbroken chain" to modern humans.

Nevertheless, the fossils point to the strong possibility that *Homo sapiens* evolved from previous hominids. All of the forms we have discussed were upright in posture; *Homo habilis* and *H. erectus* were almost certainly tool users, based on artifacts found with them. The antiquity of these forms is obvious. They existed, and they looked much more like ourselves than present-day apes. To dismiss them or take them lightly because there is not an unbroken chain linking them to us is not an honest response. The most recent forms—the Neanderthals and other sapiens—were very much like us. Neanderthals, which occurred from 125,000 to about 35,000 years before present, buried their dead and apparently cared for weaker members of their group. And fossils indistinguishable from modern human beings date back at least 30,000 years.

### *Reading the Genes*

Perhaps the most remarkable evidence of human evolution comes from comparative studies of present-day primates. The similarities in morphology between humans and the apes are obvious, and certainly are suggestive of common ancestry. Genetic studies have revealed extremely similar chromosome structure in chimpanzees and humans. For example, chimps have one more pair of chromosomes than humans; but all of the chromosomes of the two species match well in banding and size with the exception of the extra pair, which can be coupled with another pair of chromosomes to match one of the human chromosome pairs (in other words, two chimpanzee chromosomes correspond with one human chromosome). The amino acid sequence of hemoglobins of both species is identical; gorillas differ by only one amino acid in each of the two hemoglobin subunits (141 and 146 amino acids make up the two subunits of this protein).

These similarities exist in spite of the fact that the fossil evidence indicates separate evolution for at least the last 4 million years, and possibly much longer. In fact, the fossil evidence for a common ancestry of apes and hominids is lacking; none of the fossil ape-like primates dating from the Miocene epoch (24 million to 5 million years ago) is accepted as a progenitor of the hominids. Significantly, these similarities in chromosome number and gene structure would probably be considered overwhelming evidence for common descent if we were dealing with any other group of species.

### *Interpreting the Evidence*

If the chromosomes indicate such strong genetic similarity, why are we so different from the apes? Mayr argues for "mosaic evolution," where some segments of the genotype remain conservative and others change rapidly under selective pressure. What is unique about us? What has changed so rapidly? Most of the differences are considered quantitative: brain volume, hands and feet, upright posture, teeth, face and jaw, hair patterns. What about the capacity for language? Culture? Biologist Douglas Futuyma of State University of New York, Stony Brook, links these to the development of consciousness:

> The paramountly important feature of human evolution is the development of consciousness and its various manifestations: language and culture. These abilities have radically altered the course of human evolution and have brought many (but not all) aspects of human evolution almost to a halt. . . . As far as we can tell, the human capacity for culture arrived some tens of thousands of years ago at its modern level, and there has been little if any genetic change in our mental abilities since then.

Commenting on the factors that might have led up to this stage, Futuyma states:

> Just what the forces of natural selection were that directed this evolution, and what implications our primate ancestry has for

modern human behavior, are topics that have provoked an enormous amount of often quite irresponsible speculation. It is certainly possible to imagine reasons for our peculiar evolution, but probably impossible to subject any of these ideas to serious scientific testing.

By "irresponsible speculation" Futuyma is undoubtedly referring to the field of *human sociobiology*, which applies Neo-Darwinian mechanisms to human society in an attempt to establish the genetic basis for human social behavior. Sociobiologists interpret present-day human behavior by postulating the selective factors that might have been responsible for incorporating that behavior in the genetic makeup of humans during the evolutionary past. They attempt to explain the evolutionary basis for religion, morality, and in fact practically every present-day manifestation of human behavior. (In passing, we should note that human sociobiology is an extremely controversial field, one that has been subject to severe criticism on almost every claim by other biologists.)

Isn't there something more than the capacity for language and culture that separates us from the animals? Anthropologist Richard Leakey believes there is: "Humans are more than just intelligent. Our sense of justice, our need for aesthetic pleasure, our imaginative flights, and our penetrating self-awareness, all combine to create an indefinable spirit which I believe is the 'soul.'" We must go further. To understand humankind also means taking into account Auschwitz, and Hiroshima, and terrorism. How can we possibly account for the good and evil that people do, on the basis of evolutionary theory? In the last paragraph of *The Death of Adam*, John Greene wonders:

Who is to restrain man from choosing the evil which he would not do in place of the good which he would do? Is man in truth a kind of Prometheus unbound, ready and able to assume control of his own and cosmic destiny? Or is he, as the Bible represents him, a God-like creature who, having denied his creatureliness and arrogated to himself the role of Creator, contemplates his own handiwork with fear and trembling lest he reap the wages of sin, namely, death? The events of the twentieth century bear tragic witness to the realism of the Biblical portrait of man. . . . The conflict of nations and races, far from raising mankind to ever higher levels of virtue, freedom, and culture, threatens to accomplish the destruction of the human race.

Before we consider the biblical view of humankind, let us take stock:

1. The fossil record has produced evidence for several species of hominids, the Australopithecines, which lived over a 3-million-year time span without showing any evidence of change. Although upright, these were undoubtedly ape-like in behavior and intelligence.
2. Three species classified in the genus *Homo* have also left a fossil record. Two of these overlap in time with the Australopithecines; they used simple tools but were not comparable to modern humans in intelligence.
3. Modern humans appeared on the scene some tens of thousands of years ago, but to date the fossil evidence linking us to earlier forms is incomplete.
4. Modern humans show morphological and genetic similarities to the living apes, which implies common descent.
5. Modern humans possess unique characteristics that are difficult to interpret by invoking natural selection and common descent.

There are two classic—and conflicting—approaches to the origin of humankind. One is that *human beings evolved from previously existing hominids.* In accord with the previous chapter, it would be appropriate to view this process as the means God used to create humans. The other approach is special creation: *Humans were uniquely designed and created by God, in miraculous fashion.* If so, the similarities between humans, apes, and fossil forms are viewed as the result of common design. It is the conclusion of many Christians in science that the evidence is not good enough at this time to make a clear choice between these two approaches. Let us return to the biblical account to see what light the Bible can shed on these matters.

## A Different View of Humankind

Genesis 2:4 begins a second account of creation. I suggest that you read through chapters 2 and 3 of Genesis before proceeding further. As you do so, you will note that the sequence of events is different from the account in Genesis 1; the very name of God used in the account is different. Instead of simply God, we encounter the Lord God (YHWH), the name God uses in his covenant relationships with his people. This account clearly has a different intention from that of Genesis 1. Recall from chapter 5 that we considered it likely that the basic intention of the writer in the first account was to establish the doctrine of creation, revealing to God's people that he was responsible for creating them and all of the universe, and he used the seven-day framework of the account to affirm the Sabbath to the Israelites in the ordering of their lives. We should expect a different reason for the second creation account.

Once again we are confronted with questions of interpretation. What kind of literature is this? What we read is much more of a flowing story, quite in contrast to the elaborate geometric structure of the earlier account. The very fact that Genesis contains two distinct creation accounts signals to us that we should not be looking for a precise mechanism of creation as we study this passage. According to Henri Blocher:

> The style is lively and picturesque; the pictures take shape spontaneously in the reader's mind. The Lord God takes on a human form: we see him mold clay, breathe into the man's nostrils, walk in the garden

when the breeze gets up and make for the guilty couple better clothes than their improvised cloths. There is a dream-like garden with strange trees and a cunning animal who opens a conversation; you could believe you were in one of those artless legends, one of those timeless stories which are the fascination of folklore. . . . It would be difficult to achieve spontaneously the art with which the dramatic tension is built up and then resolved . . . Here we have a religious and literary achievement of the highest order. . . . Again, along with several specialists, we can recognize that the way the story is treated places it with Wisdom writings. It is a learned ingenuousness which here uses common language to set forth a unique message.

To the question of what in this second account is to be understood literally, biblical scholars are not in agreement. Blocher states that not even the most conservative scholars accept a completely literal reading. There are anthropomorphisms (God sewing clothes for Adam), and clearly symbolic meaning to parts of the account (the two trees, the serpent). On the other hand, it is not necessary (or wise) to assume that this is mythology—the product of human imagination in an effort to understand some of the fundamental realities of life. What we have is more of a *translation* of the historical events recorded, rather than a *transcription*. According to Blocher, "The presence of symbolic elements in the text in no way contradicts the historicity of its central meaning. . . . The Bible gives us too many examples of most varied styles of presenting historical facts by means of images, symbols and allegories, to prevent us from confusing the two distinct categories: language and the actual subject." Blocher considers (and I concur) that *the most important historical facts unique to this second account are the existence of Adam and Eve, and the Fall.* We will set aside the details of sorting out the symbolism, content to observe that this is a matter for deeper study, one that has received a great deal of attention from biblical scholars and goes beyond the intentions of this book.

### *The Creation of Adam and Eve*

The first account presents human beings, male and female, created in God's image; the second account gives names to God's special creation (Adam and Eve) and pictures creation in his image as a process rich with symbolism. God takes the stuff of the earth, fashions a man, and breathes life into him. He later demonstrates that humankind was meant to be both male and female—and would not be complete otherwise—by making a woman. Thus the emphasis in Genesis 2 is on the relationship between man and woman—their similarity (woman is taken from man, flesh of his flesh), and their sexuality.

Deep theological doctrines introduced here are commented on later in Scripture: the fundamental relationship between man and woman (1 Cor. 11:3, 1 Tim. 2:13), and the institution of marriage (Matt. 19:1–9). Christ himself refers to this passage in presenting his instructions on the marriage relationship. God's design—provided as a comment by the author of Genesis in 2:24—is clear: as Derek Kidner put it, "leaving before cleaving; marriage, nothing less, before intercourse." As Scripture later testifies, marriage and the sexual union that accompanies it is meant to be such a deep and lasting relationship that it is used as an analogy of God's relationship to his covenant people, and then of Christ's relationship to his church.

### *In God's Image*

The Bible says surprisingly little about what creation in God's image might mean. We might be tempted to assume that creation in God's image is what it means to be different from the animals—our rational capacity, our personhood, our capacity for language. But the term refers to a relationship first, and must be seen as a way of defining what we are in relation to God. We are, as Genesis 1 makes clear, creatures of

God who live in a world he created. But to image God means that we are, in Blocher's words, "to be the created representation of [the] Creator, and here on earth, as it were, the image of the divine Glory, the Glory which mankind both reflects and beholds."

What is it about us that images God? Surely not our physical bodies (God is a Spirit!); indeed, Gen. 2:7 symbolizes the transfer of God's image as an in-breathing by God into a body previously formed "from the dust." Imaging God is thus related to the spiritual attributes of God—not in any sense exactly, for we are only an image, but in some measure we demonstrate what God is like. We do this when we search for truth and beauty, when we are concerned about justice and other ethical issues, when we recognize the high importance of morality; these are God-like qualities.

Perhaps the best way to understand this mystery is to look to Jesus Christ, whom the apostle Paul refers to as "the image of the invisible God. . . . For God was pleased to have all his fullness dwell in him" (Col. 1:15, 19). Christ is the best representation of God, and if we would image God in our lives, then we must become like Christ. In this light, being created in God's image means that we have the possibility of becoming like him—in a reflected way, for we are still mere creatures. But think of this possibility: Although God wants us to see his glory in the creation, he has specifically created humankind to reflect his glory, in ways that the rest of creation simply could not. This is the clearest basis for human dignity, for the high value we place on human life. For this reason, the severely impaired, the senile, the most degenerate human beings deserve to be treated with respect. All humankind bears God's image to some degree; and although that image has been affected by the Fall, we are never in a position to judge that a human being is so completely devoid of humanity that he or she can be treated inhumanely.

James Houston (of Regent College, Canada) believes that it

is our capacity for sovereignty that most closely demonstrates how we image God. He has given us capabilities that equip us for sovereignty over our environment, other creatures, and over ourselves—the characteristics that make us human. But these are the consequences of being in God's image rather than the definition of it. We image him in how we exercise that sovereignty. And as we read on in Genesis, we begin to understand what it means to image God as sovereigns. We are given, in Genesis 1:28, the responsibility of ruling over his creation—of having dominion. And in Genesis 2, Adam (as our representative) is given the task of cultivating the Garden of Eden—of using and developing this part of God's creation. We will explore these responsibilities in detail in the next chapter. Psalm 8:3–6 gives us an eloquent picture of the relationship between imaging God and ruling over his creation:

> When I consider your heavens, the work of your fingers, the moon and the stars, which you have set in place, what is man that you are mindful of him, the son of man that you care for him? You made him a little lower than the heavenly beings and crowned him with glory and honor. You made him ruler over the works of your hands; you put everything under his feet . . .

The writer of Hebrews tells us that this passage also refers to Christ, in keeping with the concept that we are to be like him as we image God (Heb. 2:5–9).

### *Roots of Discord*

Genesis 2 leaves us with a paradise-like picture of a good creation, a realm where man and woman could live in harmony with each other, with the other creatures, and with their God. But as responsible creatures, we were given the freedom of choosing how we would exercise that responsibility. And as subjects of a sovereign God, we could also exercise sovereignty in our own affairs. So we encounter Genesis 3, a passage that is absolutely essential for understanding the full measure of humankind as we know it today.

Instead of attempting to deal with the question of what is symbolic and what is literal in this chapter (for fear of being submerged in a theological swamp!), we will ask a more basic question: What happened? As Blocher puts it, "Two sentences could act as a summary. . . . The author tells us of the act of revolt: the man and the woman ate of the forbidden fruit. He tells us of the deadly enticement: the man and the woman yielded to the snake."

We would like to minimize this event. After all, there is much that is symbolic—the special tree, a talking serpent, a forbidden fruit. It would have been much more impressive if Adam and Eve had gotten into a good brawl. Instead, they ate something that was not on God's approved list. But it is the unique viewpoint of the Bible that the Fall *of man* is the key to understanding all the evil and human failure in the world. This eating of a forbidden fruit is regarded as one of the most crucial acts in history. Why is that? The most straightforward translation of this first sin is that it was a revolt against the Lord God, a direct disobedience of his command. Blocher looks for the motives behind this act of rebellion, and finds *doubt* and *desire*—doubt about the goodness of a God who would forbid something that might be pleasureable, and desire for something attractive that has been forbidden.

And what about the serpent? After all, this crafty creature planted the doubt and encouraged the desire that led to the sin. This is no "snake," as we are inclined to see in the Sunday school pictures of this first sin. The subsequent curse in Genesis 3:15 clearly indicates a creature whose existence is as long as humankind's. The book of Revelation (20:2) identifies Satan as "that ancient serpent." The serpent is the Adversary, the one who was the first to rebel against God and continues down through history to oppose him and lead humankind's rebellion.

We can now see this first sin as the earthshaking event that it was. Surrounded by all the good things of God's creation, in constant fellowship with the Creator and in harmony with his commands, humankind listens to a lie, and does the will of the liar. The fruit is good, and in taking it, Adam and Eve pervert it into evil use. Blocher's comments on this are very perceptive:

> Thus, it is always with regard to one of the fruits of God's garden, a fruit that is genuinely beautiful, pleasant and useful, that mankind is tempted. Evil is not in the good that God has created, but in the rejection of the order that God has instituted for the enjoyment of the world. Temptation plays with the facets of things that are good, and highlights the attractions of the beauties in creation. Sin then perverts the excitement which these objects quite rightly cause within us. Thus, to revert to John's words, "the lust of the flesh" perverts and corrupts the excitement which drives us towards what is good and beneficial. The "lust of the eyes" likewise corrupts the drive towards what is beautiful and true. The "pride of life" perverts the rightful effort to be, and to be valued.

Let us admit that evil and this concept of an "original sin" are mysteries. Their origin, shrouded in the past and clothed with symbolism, is far less important than their effects. Two of the effects stand out: the effect of sin on our ability to image God, and the effect of sin on the rest of God's creation. Although we continue to enjoy the privilege of being in God's image, that image is constantly impaired by our sinfulness; the warfare within us is thoroughly described in Paul's writings (see Romans 7). And Scripture is also clear in stressing that the Fall has affected creation. God's good creation has been corrupted—the creatures and the environment that so wonderfully support human life and activity have suffered.

As a result of the Fall, the God-given potential that we have to develop human culture has been twisted. The presence of sin has brought a new dimension into the world. There are new expressions of the things of creation that were never part of God's handiwork: The goodness of sexuality is perverted into prostitution; the ordained character of the state is twisted into anarchy or dictatorship; natural chemical substances are extracted and used to alter and enslave people's minds. It is fair to say that most of the remainder of this book is an excursion into the aftermath of the Fall as it affects the living world. It would be hard, therefore, to overemphasize its significance. It is also important to underline its historicity.

### The Historicity Question: Who Were Adam and Eve?

We have traced two views of the origin of humankind. Do they converge? To be consistent with the spirit of this book, we should expect them to. Several New Testament passages reaffirm that Adam and Eve were the first humans (Luke 3:23–38; Acts 17:26; Rom. 5:12–21). May I suggest that it is entirely reasonable to add, the first humans to bear God's image? And what of the fall? Where did that happen, and exactly how? Did God create Adam *ex nihilo* (literally, from nothing)? Or did he use preexisting matter (dust?) in hominid form? Exactly when in history might this have happened? Were the Neanderthals really humans (in God's image)? These are questions for which there is far too little evidence to venture an answer. They are what we shall have to leave as loose ends, even though there are many who would like to see them tied.

As we mentioned earlier, there are two classical views on the origin of humankind: Humans either evolved from preexisting hominids, or else they were created by God with many features that are similar to fossil and existing primates. The former explanation rests comfortably within the scientific framework, although on less data than many would like to see. The latter explanation—special creation—cannot be dealt with by science. Both of these views are held by different groups of Christians who believe in the reliability of the Bible.

And there are some who believe that God in a miraculous way transformed preexisting hominids into the first humans, a sort of middle-of-the-road view. As I suggested earlier, the evidence simply is not capable of distinguishing clearly between these views. Accordingly, let me suggest that we should not make these questions about human origins a test of orthodoxy. And let me also suggest that we should not allow this question to be a barrier to faith.

Henri Blocher deserves a last word here:

> It is perhaps salutary for us thus to end on an inconclusive note, which in certain respects illustrates the attitude of faith. For faith does not have all the answers straight away. Nor does it claim that contemporary science gives it complete support. If certain factors in today's scientific picture appear contrary to the Word of God, faith is not shaken. It has such confidence in that Word that it can be quite open about its hesitations and wait patiently for the clouds to clear.

## Summary

Darwin's work opened up the question of human evolution, which seemed to threaten belief in the Bible and became steeped in controversy. At the time, there was little evidence in support of human evolution, but continuing research in anthropology and biology has turned up evidence that must be confronted.

The fossil record has yielded hominid forms that have been classified into two genera (*Australopithecus* and *Homo*), each of which contains several species that date back several millions of years. Although some publications refer to an unbroken chain of species linking the earliest forms to modern humans, most working anthropologists admit to the existence of many gaps and uncertain relationships in the fossils. Other evidence consists of the many anatomical and genetic similarities of humans with present-day primates. However, modern people possess unique characteristics that are difficult to interpret according to ordinary evolutionary theory. Current approaches to the origin of humans consist mainly of the view that humans evolved from preexisting forms (theistically, as God's creative method), or the view that humans were uniquely and miraculously designed and created by God. The evidence is not yet good enough to make a choice between these two views.

The second account of creation, in Genesis 2 and 3, focuses on the creation of humans. Differences from the Genesis 1 account indicate a different purpose for this passage, and Henri Blocher refers to it as more of a translation of historical events than a transcription. Symbolic elements are clearly present, but two historical facts unique to this account are the existence of Adam and Eve, and the Fall. The second account emphasizes the relationship between man and woman and involves the concept of creation in God's image, introducing doctrines that are developed richly later in the Scriptures.

The Fall of Adam and Eve into sin, recorded in Genesis 3, contains much symbolism but is a key to understanding most of human and biblical history. In this act of rebellion, the first humans listen to a lie and, doubting God, they do the will of the lier, who is Satan. The effects of the Fall extend both to our ability to image God—it is now marred—and to the rest of creation, as our sin now corrupts everything that we do. The rest of this text deals with the effects of the Fall.

Many questions remain as we face the mysteries surrounding our own origin. The evidence simply is not good enough to distinguish between the various views held by different Christians. We should not make these questions about our origins a test of orthodoxy, neither should we allow this problem to be a barrier to faith.

*Chapter 9*

# STEWARDS OF CREATION

> The rule of no realm is mine, neither of Gondor nor any other, great or small. But all worthy things that are in peril as the world now stands, those are my care. And for my part, I shall not wholly fail of my task, though Gondor should perish, if anything passes through this night that can still grow fair or bear fruit and flower again in days to come. For I also am a steward.
>
> GANDALF, IN *THE RETURN OF THE KING,* BY J. R. R. TOLKIEN

## Introduction

The wizard Gandalf is a central figure in J. R. R. Tolkien's classic trilogy, *The Lord of the Rings.* The quote above comes from a scene where Gandalf is confronting Denethor, the Steward of Gondor. Denethor senses that doom is coming upon his realm as the forces of evil grow stronger and march in war towards Gondor. Yet he is a proud man and, in the end, his pride leads him to downfall as he seeks hidden knowledge of the enemy's purposes and becomes trapped by the enemy's overpowering will. Although he is a steward of his kingdom, Denethor sees his stewardship as an impediment to the elevation of his own family to complete rule of the kingdom. Gandalf urges him to action, even though there seems to be little hope that the forces of good can prevail. But Denethor's pride and selfish desires and his unwise contacts with the enemy so corrupt his purpose that he loses his mind and perishes in self-inflicted flames.

There is a stark contrast between these two stewards—Gandalf and Denethor—that speaks to our present situation. Gandalf feels responsible for the survival of "all worthy things"; as the story closes, the success of his stewardly care actually leads to the passing of his own age, middle earth. Denethor's "stewardship" is no stewardship at all, for his concern is primarily for his own designs on the kingship; in the end, his pride consumes him.

If we take a realistic picture of humankind's rule over the earth, we can see the evidence of both good and poor stewardship. On one hand, many of the diseases and parasites that once took a terrible toll on human life are well understood, and life expectancy continues to increase in most nations. More people are being fed than ever before and, in many countries, with far better nutrition. Research into ecosystem functioning has given us a growing understanding of how natural systems work. Millions of acres of natural areas have been set aside for preservation of wildlife and scenic beauty. Regulations have led to significant reductions in pollution in some of our cities and waterways. These things are the results of good stewardship.

But the evidence of poor stewardship surrounds us. As our civilization anticipates crossing the threshold into a new century, we are faced with a battery of problems with global proportions. The human population surpasses 5 billion and continues to increase exponentially. The gap between rich and poor nations in wealth and resource use is increasing. We are approaching the limits of the capacity of biological systems to produce food and fiber in support of human life and economic activity—in some parts of the world, those limits have been exceeded. The destruction of tropical rainforests and the human intrusions into other natural ecosystems are causing both the loss of genetic diversity and a high rate of extinction of the nonhuman inhabitants of the planet. Our technology has created the acid rain that is destroying lakes and forests, and gives off the increasing amounts of carbon dioxide that will inevitably change global climate and bring about a rise in sea level. Like Gondor, our realm is threatened; the potential for global disaster exists, not only in our arsenal of nuclear weapons, but also in our misuse of the land, air, and waters that support life.

## An Indictment

These problems will not simply work themselves out in the normal course of events. In fact, it has been the pursuit of "business as usual" that has led us into our present plight. We are not living wisely; our species is not in any sense at equilibrium with our environment. There is something fundamentally at fault in our basic approach to the natural world. What we do to the environment is usually regulated by law, but our laws clearly allow us to do too much—and our economics tell us that we have no choice. We must ask if there might not be something seriously wrong with our economic and legal and ethical systems that have allowed us to proceed so far along the road to resource depletion, overpopulation, species extinction, and pollution. This is obviously not a trivial pursuit—if we can finger the culprits, the flaws in our systems, we might learn how to cope better in the future.

It might come as something of a surprise to you that some very influential writers have pointed the finger at Christendom. The Judaeo-Christian teachings concerning humankind and nature are at the root of our environmental problems, according to these writers. In a significant book, *Design With Nature*, Ian McHarg writes:

> The great western religions born of monotheism have been the major source of our moral attitudes. It is from them that we have developed the preoccupation with the uniqueness of man, with justice and compassion. On the subject of man-nature, however, the Biblical creation story of the first chapter of Genesis, the source of the most generally accepted description of man's role and powers, not only fails to correspond to reality as we observe it, but in its insistence upon dominion and subjugation of nature, encourages the most exploitive and destructive instincts in man rather than those that are deferential and creative.

Historian Arnold Toynbee traced the origins of pollution to one Bible verse, Genesis 1:28, where we read, "God blessed them and said to them, 'Be fruitful and increase in number; fill the earth and subdue it. Rule over the fish of the sea and the birds of the air and over every living creature that moves on the ground.'"

The most widely known indictment of Christianity came from University of California, Los Angeles, historian Lynn White, Jr., in a much-republished essay entitled "The Historical Roots of Our Ecologic Crisis." According to White, these roots lie in the Judaeo-Christian teachings that humanity, created in God's image, was set apart from nature and encouraged to investigate and exploit it. This led to the eventual development of science and technology in the Western cultures, where Christianity was so influential. In White's words:

> We would seem to be headed toward conclusions unpalatable to many Christians. Since both science and technology are blessed words in our contemporary vocabulary, some may be happy at the notions . . . that . . . modern science is an extrapolation of natural theology and . . . that modern technology is . . . [a] . . . realization of the Christian dogma of man's transcendence of, and rightful mastery over, nature. But, as we now recognize, somewhat over a century ago science and technology—hitherto quite separate activities—joined to give mankind powers which, to judge by many of the ecologic effects, are out of control. If so, Christianity bears a huge burden of guilt.

In his perceptive book A *Worldly Spirituality*, Wesley Granberg-Michaelson has added an additional accusation. He speaks of a strong element of dualism within evangelical Christianity, of a division into spiritual and material realms, where the soul and spiritual concerns are what really matter and are separate from the world. The earth and material things will all pass away. In fact, we should not be surprised that things are getting worse; these are indications that Christ's second coming is at hand. Accordingly, there is no point in trying to prevent the downward spiral of resource depletion and ecological ruin.

So not only is Christianity considered to bear responsibility for the worldwide destruction and exploitation of nature, certain elements within the Christian church actually welcome these developments as a sign that Christ's return is near. These charges require a response; they deal with serious issues. I will first examine our current attitudes toward nature, and then after tracing the history of exploitation I will give my conclusion as to the roots of our ecological problems. Following that, I would like to turn to Scripture to explore the themes of dominion and stewardship. I will show you that far from being the source of our environmental problems, Christian faith provides a redeeming vision for the future and a path for reaching that vision.

### *Response 1: Our Current View of Nature*

Our society exhibits a kind of schizophrenic relationship to the natural world. On the one hand, we sentimentalize it. Images of natural settings are associated in advertising with all sorts of unrelated products. We flee from our urban homes to parks and forests for vacations, but most of our contact with nature is through the automobile window. Our cities and malls and streets bear the names of the natural settings they have replaced.

On the other hand, we dominate nature. Land represents a commodity to be exchanged and used. The term "natural resources" indicates our approach to the elements of nature. If we conserve parts of nature, it is either for our future use or for our present benefit. We have the objective of bringing nature under control, harnessing it for our economic good, and enclosing it for our protection.

Granberg-Michaelson suggests that this attitude is a uniquely modern development brought about by a fundamental change in our ideas of how God is related to nature. We have, in effect, secularized nature—removed it from any connection to God and therefore changed the terms of human relationships to the natural world. In his words:

> Nature is something "out there," apart from us and apart from God. This detachment enables first of all objective study, and then manipulation and control. Domination and exploitation of this external material of nature can be undertaken without any religious restraints. Meanwhile, a superficial and nostalgic picture of nature is fostered by advertising images and preserved in wilderness areas and wildlife preserves.

In support of this thesis, Granberg-Michaelson traces the roots of our modern attitudes toward nature to the Enlightenment and the Scientific Revolution, and then shows how subsequent developments in human thought led to a gradual removal of God from major elements of human culture. He summarizes as follows: "The modern mindset, in its understanding of the world, of nature, of reality, of history, of self, has peeled away the assumption of God's existence, or God's relationship to anything that matters." This is in fact the essence of the worldview known as *secularism*; God has been pushed out of the picture.

Freed from the operation and even the traces of religious restraint, we may now indulge in the pursuit of materialistic goals, using the natural world as the means of achieving those goals. And so another element of the modern worldview—*materialism*—also appears as a consequence of our alienation from God. The attractions of wealth, luxury, and sensual indulgence so efficiently manipulated by the advertising media lead us inevitably to greater consumption, and this in turn results in greater environmental impact (to be explored in chapter 12). So Granberg-Michaelson concludes that rather than Christianity, it has been the emergence of the modern worldview and its freedom from any religious restraint in pursuit of materialistic goals that is responsible for the exploitation of nature and its consequences. But there is an even more basic root for these problems, and we can trace it by considering the history of exploitation.

### *Response 2: The Historical Record of Exploitation*

According to McHarg and Toynbee, the Genesis command to humankind to have dominion and subdue the earth has encouraged the exploitation and misuse of nature by cultures influenced by biblical religions. This thesis can be put to a historical test. If it is true, civilizations developing outside of Western and Judaeo-Christian influences should reveal humankind living in harmony with nature and not exploiting it. One such civilization is mainland China. Geographer Yi-Fu Tuan has documented the centuries-old impact of Chinese civilization on the landscape, citing extensive overgrazing and forest removal over vast areas and subsequent major soil erosion. This has occurred in spite of centuries of Taoist and Buddhist teachings about humankind as part of nature. Commenting on this discrepancy, Tuan states: "A culture's publicized ethos about its environment seldom covers more than a fraction of the total range of its attitudes and practices pertaining to that environment. In the play of forces that govern the world, esthetic and religious ideals rarely have a major role."

It is a fact that the skeletons of many dead civilizations are spread around the world—Romans, Greeks, Assyrians, Incas, Mayans, Khmers—and the evidence of their environmental misuse can be seen in the harsh and sterile landscapes once occupied by these civilizations. A search of the Old Testament does not show any evidence in Hebrew technology of unique exploitive practices. Indeed, the Egyptians, with their nature gods, developed a much higher technology. Clearly, exploitation and misuse of nature can be seen in many civilizations and human societies, not just those influenced by Judaeo-Christian beliefs. Humankind's dominion over nature is a fundamental human characteristic, one that we traced in the last chapter to God's creative action.

Our technology—the attitude toward nature or objects in nature that sees them as something to be put to use to serve human needs—is not therefore the consequence of our efforts to obey God's command in Genesis 1:28. It is the outcome of capabilities given to our species. In a sense, human technology is comparable to the way all species make use of their environment and other species to meet their requirements. Beavers cut down trees, build dams, and flood lowlands; gypsy moths and budworms defoliate vast areas of forests. However, human use of stored energy and machines has multiplied our power over creation well beyond sheer biological "muscle power."

### *The Heart of the Problem*

There is, of course, a vast difference between the way animals and people make use of their environment. That difference relates to the great gulf separating humankind from other animals. Not only are we much more capable of manipulating our environment, there is a moral dimension to what we do. We can choose. We have the ability to act wisely, or otherwise. We can do good, or evil. We can do miraculous things, and monstrous things. And evil enters when technology coupled with human carelessness and pride leads to environmental breakdown, when the use of natural resources coupled with greed becomes exploitation.

There is no need to search further to find the common denominator for human exploitation and misuse of nature. The explanation reveals itself every day, if we care to look for it. It is present in each of us: *human greed, pride, carelessness, and ignorance.* This is the source, the root of our environmental troubles. In our Fall and continued rebellion against God, we exploit, destroy, and otherwise misuse the capabilities and great gifts God gave to us. Toynbee, White, and McHarg were impressed by the Genesis dominion mandate; they should have read further. They did not acknowledge the Fall.

So if we want to find direction for coping with the serious environmental problems that confront us, we must recognize human nature for what it is and be willing to deal effectively with it. Science and technology cannot lead the way—they are the servants of a modern worldview that has people instead of God at its center. Christendom is not responsible for our problems. To lay the blame for them on Judaeo-Christian beliefs is to misread history. The damage that this accusation may do is not in discrediting Christianity—our faith will certainly survive these charges. But in convincing some that the accusation is true, this charge misdirects the emphasis for action. White, Toynbee, and others suggested that we need to look to other religious traditions, ones that emphasize human beings as part of nature. I would argue that what we really need—at the level of our whole society—is a recognition of the realities of human nature as they are revealed in Scripture, and a willingness to deal with them in our system of laws, education, and other human institutions. We need good stewards, wise managers of natural resources, thoughtful caretakers of the wild creatures that share this planet with us, people who love the creation. And at the level of Christendom, we need a rediscovery of the responsibilities toward nature that Scripture actually teaches. To that topic we now turn.

## The View from Scripture

### *Dominion as God's Image-bearers*

The testimony of Scripture is clear: God has given humankind dominion, or rulership, over his creation (Gen. 1:28; Gen. 9:1–7; Psalm 8:4–8). However, our authority is a limited, derived authority, and is meant to reflect the ultimate authority of God. We represent him on earth, the only creatures that are able to do so. We are to image God in our rule, doing so in a way that demonstrates what God is like. This is clearly not a license to exploit the creation—quite the opposite. Our calling is to act as God's agents, or stewards, and this passage is rightly seen as the root of biblical stewardship.

Genesis 1:28 also speaks of "subduing the earth," which has traditionally been given a somewhat different meaning from "ruling." It is seen as a fundamental application of the task of ruling. The immediate object is the earth—meaning the land, not the "world"—and the clearest intent of the passage is to cultivate the ground for food, humanity's most essential need. The second creation account repeats this emphasis: Adam and Eve are placed in the Garden of Eden to work it and take care of it. They are also given the responsibility of naming the animals, which in the ancient Hebrew culture implied the ability to learn their inner secrets, to know them thoroughly—an exercise of human intelligence.

These tasks—subduing, having dominion, cultivating—all point to the development of a culture. And since all of this is God's clear intention for humankind, the term *cultural mandate* has been used to describe this most basic of human responsibilities before God. We have a mandate from God to be the cultivators of the good things of his creation. This mandate means that God has intended for humankind to interact with his creation in such a way that we would develop a culture. In doing so, we use the created elements and so demonstrate clearly our dominion.

In the same verse in which we read about dominion, we also read that God commands humankind to "be fruitful and multiply," an indication that it would take many future generations to carry out the cultural mandate. And early Genesis goes on to document a few examples of the development of culture: agriculture, the domestication of animals, the invention of musical instruments, the forging of metal tools.

Culture is wrapped up in human history, the record of what we have developed over time. It covers the whole range of human society—our music, art, technology, economic and political institutions, education, media, and so forth. It is no exaggeration to claim that to be human is to be a cultural being. It also seems clear that this development of a culture was part of God's original plan for the world, that creation was meant to be developed. But there is another side to Adam's task.

### *Serving the Creation*

For Adam and Eve, the Garden of Eden was all of creation. God instructed them to *work* and *take care of* the Garden. Both of these words in Hebrew suggest a deferential attitude toward the object (the creation), an attitude of preservation and service. This attitude can be seen as a strong counter to the more forceful expression of dominion in Genesis 1—subdue and rule. It restricts the ways in which our dominion may be exercised. We are not the owners of creation but the stewards, masters but not slave-masters. God is the model for how we are to carry out dominion; we are his image-bearers. Granberg-Michaelson expresses this well:

> Serving and preserving the creation is rooted in the orientation of one's life to God. The view of Genesis is that the Lord blesses life, sets forth the values for life, and issues the promises for securing life—both for humanity and all creation. Life lived in relationship to the Lord God necessarily and naturally is a life that participates in sustaining the creation.

The significance of this task is related to the value and purpose of creation, a theme we explored in chapter 2. The creation brings glory to God; its goodness speaks of his goodness, its beauty of his beauty. Adam's task of caring for the creation must be seen in this light.

### *The Fall: An Impact Statement*

As we know, the story takes a turn downward from here. Genesis 3 records the act of rebellion that we call the Fall. Our representatives, Adam and Eve, were led by Satan into disobedience. Believing the lie that they would be more like God if they ate of the fruit, Adam and Eve achieved just the opposite. Their disobedience brought discord and alienation into God's good creation. This alienation intruded into every human relationship—between humans and God, among humans, and between humans and the rest of creation.

Sin also brought judgment. Nature—the creation—suffered as a consequence of the Fall. One effect, symbolic of the alienation between humankind and the creation, is recorded in Genesis 3:17 and 18: "Cursed is the ground because of you; through painful toil you will eat of it all the days of your life. It will produce thorns and thistles for you . . ." Some scholars suggest that a better translation of the Hebrew would be: "Cursed is the ground to you . . ." indicating that God's judgment falls primarily on our efforts to use the creatures and the land for our purposes. There is no suggestion here that the creation was morally degraded as a result of human sin. Indeed, Scripture teaches that creation continues to tell of God's glory (see chapter 2).

In effect, this curse means that our cultural task will now be a struggle. We will often view the earth as an enemy, and it will bear the battle scars of our assaults. The very culture that we develop will also show the impact of the Fall; two forms of its expression will appear, reflecting the two kingdoms that are now at war—God's kingdom and Satan's. The entire range of human activities becomes a battleground, for there is nothing in the creation or culture that remains unaffected by sin. Our dominion will no longer reflect clearly the rule of God's image-bearers, for that image is no longer an accurate reflection of God. Indeed, all that we do is affected by the fall. We pursue the development of culture without obeying God's laws, and we rule over creation as if we were the owners.

Misuse of human dominion can be traced throughout the Bible. Early chapters of Genesis record a crescendo of human violence that culminated in the judgment of the flood. Noah and his family represent one of the few examples of the proper exercise of dominion, as they skillfully fashioned a vessel to preserve the creatures in the face of coming judgment. But then we read in Genesis 11 of the Tower of Babel, a symbol of the resurgence of human pride, and God's judgment which followed. The book *Earthkeeping* summarizes:

> From Abraham (who is asked to sacrifice his son, his one link with dominion) on through the prophets (who continually proclaim to the people God's message that their strength is not in their chariots, but rather in their giving up of their strength to God), the Old Testament is a record of this painful lesson which God must teach his people: whether we speak of humanity or nature, dominion does not mean simply imposing one's will on the weaker.

Thankfully, where there is alienation, rebellion, and judgment, there is also the possibility of reconciliation and redemption. The fall of humanity is only part of the biblical message, and the Gospel is the good news that God has made a way for the healing of broken relationships. An interesting passage in Romans speaks of this healing for the creation: "The creation waits in eager expectation for the sons of God to be revealed . . . the creation itself will be liberated from its bondage to decay and brought into the glorious freedom of the children of God . . . the whole creation has been groaning . . . right up to the present time . . ." (Rom. 8:19–22). Our dominion and sin have brought evil consequences on the rest of the created world. But as this passage suggests, the destinies of humanity and the creation are tied together in a redemptive way. We need now to explore God's intention for the exercise of dominion as it is expressed in the concept of stewardship.

### *The Biblical Concept of Stewardship*

"To the Lord your God belong the heavens, even the highest heavens, the earth and everything in it" (Deut. 10:14). Just in case the Israelites weren't clear about this, Moses reminded them who was really in charge. The theme of God's ownership of the creation is repeated in the Psalms: "The earth is the Lord's, and everything in it, the world, and all who live in it" (Psalm 24:1; see also Psalm 50:10–12). Our dominion is a delegated one; and, as we have seen, it involves both developing a culture and caring for the creation. We are to do this as God's image-bearers and also his servants. Now a servant who is put in charge of his master's property is a steward. This concept—stewardship— captures in one word our proper relationship to God's creation. We are managers, but not owners. We may put God's property to use, but not misuse. As stewards, we are managers of God's household "for the welfare of the creation and the glory of God," as *Earthkeeping* records it. So we are to serve the creation and care for it.

To the list of stewardship responsibilities we must add accountability. We are responsible for the way in which we exercise our stewardship—for the uses to which we put the creation, and for the way in which we preserve it. We alone among the created beings have been singled out for this task. And it is important to recognize that God has made us enough like himself so that this task is not beyond our capabilities. He has also revealed himself to us in his Word so that we can know and accept and pursue his priorities.

It follows that as God's stewards we must understand what it is we are to manage and serve. The information is already out there in the creation, and it is our culture-forming task to gather it in and then act on what we have learned as we seek to be good stewards. But knowledge and the desire to use it wisely are not enough. We must also *love* the creation, and the real source of that love is in our love for the Creator. As 1 Corinthians 13 makes clear, if we understand all mysteries and all knowledge, but do not have love, we are nothing. We must be engaged with the creation at the level of our hearts. This great creation is not hard to love; on the contrary, many outside the Christian faith express a deep love for nature. Should not we who know the Creator find it even easier to love what he has made? Is it not absolutely sinful that we who worship God so shut ourselves off from the creation in our cities and suburbs that we cannot even see God's works?

Stewardship, then, includes the following meanings:

1. Our dominion is a derived authority; God maintains his ownership.
2. As stewards, we serve the creation; we have a concern for its welfare.
3. We are accountable to God for the conduct of our stewardship; we are people to whom a trust has been given.
4. We need to understand how the creation works, and then act wisely on that knowledge.
5. We must love the creation, even as we love the Creator. When we do, our responsibility and knowledge and action will connect with the heart, and only when this happens will we be capable of acting in God's place.

### *New Testament Themes*

Granberg-Michaelson in *A Worldly Spirituality* relates how his early years as a Christian were directed toward "personal salvation." He read the Old and New Testaments primarily as a guide to building a deeper personal faith in Christ. But there came a time when his awareness of injustice, violence, and environmental problems caused him to begin, in his words, "wondering if faith in Jesus Christ was only a personal matter."

And he began a journey that led him to a more outwardly active faith and to a renewed appreciation of the broad scope of biblical truth. One important outcome of his journey is his recording of New Testament themes concerning Creation, humanity, and God. I would like to highlight two of these themes as they relate to the understanding of the stewardship of creation that I am trying to develop. And in passing, I would also like to commend to you the journey taken by Granberg-Michaelson. It is a journey that, if enough Christians join in, can have great meaning for our stewardship tasks.

One important theme of the last chapter is that humanity's creation in God's image is restored and demonstrated in its fullness in the life of Jesus Christ. "Christ, who is the image of God. . . . For God, who said, 'Let light shine out of darkness,' made his light shine in our hearts to give us the light of the knowledge of the glory of God in the face of Christ" (2Cor. 4:4,6). It is in his servanthood that Christ demonstrates to us the shape that our dominion and stewardship should take. We are told in Philippians 2:5–8 to have the attitude of Christ Jesus, who took the form of a servant, and "humbled himself and became obedient to death—even to death on a cross." Our posture, then, as we become imitators of Christ, is one of servanthood toward all of creation.

Another important New Testament theme centers on the redemptive work of Christ and its importance for all of creation. John 3:16 tells us that "God so loved the world [literally, the cosmos] that he gave his one and only Son . . ." The purpose of his coming was to restore the relationships broken by the Fall—and this restoration, or reconciliation, extends to all of creation, as we read in Colossians 1:20: "Through him [Christ] to reconcile to himself all things, whether things on earth or things in heaven, by making peace through his blood, shed on the cross." In Granberg-Michaelson's words, "The power of sin seeks to destroy humanity's fellowship with God, breed enmity among human beings, alienate humanity from creation, and infect creation with disorder and violence. The power of God's redemption in Christ Jesus restores fellowship between creation, humanity and God."

### The Challenge of Stewardship

Redemption brings a new significance to stewardship. Let us summarize the biblical themes we have discussed: Humanity was given dominion and appointed as stewards over God's good creation; this involved developing human culture and caring for the creation. However, our early representatives were led by Satan into disobedience, and brought alienation and God's judgment as a consequence. Every human activity from then till now shows the effects of sin; and, in particular, this extends to the creation. God's good creation has suffered as we have misused our dominion to exploit and ravage it. But God has not abandoned his creation. He sent his Son to redeem his fallen creatures and restore creation's goodness. This involves the culture we have developed, as well as the natural world that we live in. Our stewardly task is therefore extended beyond basic management and wise use; it now involves an opportunity to participate in this redemption by bringing healing to the creation and the restoration of goodness to the culture. And we do this out of hearts filled with love for the Creator and his creation.

In this light, it is now easier to see the significance of the dualism in evangelical Christianity referred to earlier in the chapter. It is wrong to attempt to divide our world into spiritual and material realms. A thoroughly biblical worldview sees the world as one creation. This world does matter; it is God's creation, and he calls us to be involved in its redemption. We may not stand aside and watch things get worse as we abandon the world to human greed, misuse, and exploitation while hoping for the soon return of Christ. In Chapter 12, we will examine the current set of problems that I have called the environmental revolution and bring dominion up to date. And in the final chapter of this book, we will explore our Christian task in more detail. However, we have not yet dealt with one important question: *Is stewardship only a role for Christians?*

### *The Cyrus Principle*

I wish at this point that I could direct your attention to the strong leadership of the Christian churches in stewardship concerns. I wish I could say that many Christians are on the cutting edge of society's attempts to cope with and solve environmental and resource issues. Sadly, more often than not, the Christians are right in the middle of the unruly mob that we have become as we continue to use resources as if there were no tomorrow, as we spread wastes over the face of the earth and into the air. The message of stewardship is often a muted one, not often proclaimed from the pulpit or practiced as an outworking of faith. We have become far too comfortable with the dominant Western worldview that has rejected God and now serves other "gods," pursuing material goals in the hope of achieving happiness and fulfillment. Our way of life more faithfully reflects the culture around us than it does the gospel. It would seem that if stewardship is dependent only on the work of the Christian church, we are doomed.

It is good news for the creation and its future that many people who do not recognize themselves as stewards of God's creation nevertheless are carrying out their dominion in a thoroughly "Christian" manner. Dominion belongs to all humanity, and so all wise and stewardly exercise of that dominion constitutes obedience to God's norms, regardless of whether God is acknowledged.

Consider Isaiah 45 in the light of this observation:

This is what the Lord says to his anointed, to Cyrus, whose right hand I take hold of . . . "I call you by name and bestow on you a title of honor, though you do not acknowledge me. I am the Lord, and there is no other; apart from me there is no God. I will strengthen you, though you have not acknowledged me . . . I will raise up Cyrus in my righteousness; I will make all his ways straight. He will rebuild my city and set my exiles free, but not for a price or a reward," says the Lord Almighty. (Isaiah 45:1,4,5,13)

At a crucial point in Hebrew history, God used a pagan king, Cyrus of Persia, to accomplish his purposes. Cyrus issued a proclamation throughout his kingdom that allowed Israelites there in captivity to return to Jerusalem and build a temple for the worship of God. Cyrus may not have worshiped God, may well have given his allegiance to false gods, but Cyrus served the Lord.

Earlier in this chapter, I suggested that holding Christianity responsible for the serious environmental problems that confront us is not a reasonable reading of history. The suggestion from White and Toynbee that we look to other religious traditions for help puts the emphasis for action in the wrong arena. We need a strategy that cuts across religious and other ideological barriers, one that recognizes the true roots of our problems and deals with them as the outcome of human pride, greed, and ignorance. Stewardship begins here and works outward from this context. We need a full measure of understanding; we cannot manage well what we don't understand. We need to know how natural systems work, and how they support human life—environmental science. We must understand how the market system works, and how goods and services are exchanged—economics. We need to understand how human societies are structured—sociology. And we need to know how to get things done in human systems of government—political studies. Christians, above all, should be involved in this process—we worship the God who appoints us his stewards. But we must recognize and welcome—and often follow—the Cyruses whom God is using in often powerful and effective ways to demonstrate the meaning of stewardship of his creation.

## SUMMARY

A realistic view of humanity's rule over the earth reveals clear evidence of both good and poor stewardship. However, the current situation of overpopulation, misuse of resources, and global pollution indicates that we as a species are not at equilibrium with our environment. Something is wrong.

Several writers—McHarg, Toynbee, White—claim that Christian belief is at the root of the ecological crisis. Indeed, some within the Christian church welcome the downward trends as a sign of Christ's soon return. These charges need answering.

Current attitudes toward nature can readily be traced to the widespread secularism that has removed God from the modern worldview and allowed full pursuit of materialistic goals, with the result that nature is exploited without restraint. A look at history reveals that technology leading to exploitation and misuse of nature is common to all societies, not just those influenced by Judaeo-Christian beliefs. Indeed, human technology is comparable to the way all species make use of their environments to meet needs.

However, there is a moral dimension to human use of the environment. We can do both good and evil. The true source of human exploitation and misuse of nature is the Fall, and its outcome in human greed, pride, carelessness, and ignorance. Societies must deal with these realities of human nature and promote good stewardship; Christians must rediscover the scriptural truths concerning our responsibilities toward the creation.

The Bible teaches that although we have dominion over nature, we have a dual calling both to develop a culture and to act as God's stewards over the creation. However, the entry of sin through the Fall has made a struggle of our culture-forming task, and has led to alienation from and misuse of the creation. But where there is alienation there is also the possibility of reconciliation, and God calls us to participate in the redemption of his world. As God's stewards, we are accountable to him for the conduct of our task; we also must learn how the creation works, and act wisely on that knowledge. And we must love the creation, even as we love the Creator.

It is clear that we may not, as Christians, stand aside and watch things get worse; we are called to be redemptively involved in bringing healing to the creation. Unfortunately, Christians do not seem to be aware of their stewardly tasks, and live lives that more faithfully reflect the dominant worldview of our culture than the gospel. But there are many outside the faith whom God is using to carry out stewardly dominion—the Cyrus principle. We should welcome and join with them in these crucial times.

*Chapter 10*

# THE BIOMEDICAL REVOLUTION

If ethical principles deny our right to do evil in order that good may come, are we justified in doing good when the foreseeable consequences are evil?

A. V. HILL

## The Ethical Dilemma of Science

Knowledge brings power, and biological knowledge leads to power over the living world. From earliest times, humans have sought knowledge of their own infirmities, an enterprise that has gathered momentum until it has become the enormous fund of knowledge and battery of technological achievements that is biomedicine. It is the basis of the health care industry, the largest employer in the United States. Many who read this book are already on a track leading to professional involvement in this industry, and all of us are recipients of its services. An aura surrounds the whole field of medicine, and in particular the role of the medical doctor, that borders on worship. These are today's miracle workers; they save our lives and restore us to health, and we are grateful for their services. In keeping with the previous chapter, we are right to point to the accumulation of biomedical knowledge as a proper exercise of our dominion. We are in the best sense "subduing the earth" when we develop and use our knowledge to alleviate suffering and prevent death and disease.

But there are consequences. The application of medical knowledge has contributed substantially to lower death rates practically everywhere in the world—an entirely welcome and intended development. However, because birth rates in many nations have remained high while death rates declined, vast increases in the human population have resulted. Rapid population growth has led to serious consequences for natural resources; some parts of the world have undergone such change that they have lost their capability to support the human society that has been dependent on them for food and fiber. This situation represents one of a class of applications of science with undesirable consequences, referred to by A. V. Hill in his classical essay *The Ethical Dilemma of Science*. The dilemma is especially acute when the effects of the application are foreseeable.

Hill cites other examples of this dilemma. The most obvious one is the acquisition of knowledge in nuclear physics, which has led to great benefits to medicine and biological research and to the development of nuclear power—as well as to the stockpiling of nuclear weapons with their potential of obliterating life on this planet. Pesticides were developed in order to kill insects that eat our crops and spread serious diseases—but they also disrupted natural food chains and decimated populations of predatory birds like our national symbol, the bald eagle. Knowledge of the microorganisms that cause most of the communicable diseases that have plagued us throughout history has reduced most of these diseases to trivialities—and also made it possible to develop biological warfare and to field large armies without fear of epidemics. Indeed, many areas of scientific knowledge have been developed and applied from motives that are beyond reproach, and have, often simultaneously, led to problems with serious and often global consequences. Sometimes the problems were unanticipated. But in many cases, those who have been engaged in the development of knowledge—the scientists and technologists—were well aware that their knowledge had the potential for misuse.

The field of biomedicine has tended to raise some extremely difficult and controversial moral issues as it has accumulated knowledge. The knowledge is particularly troublesome because it often seems to have the potential to cut both ways. Some wonder if we might not have reached the point where we are forced to say that some areas may be investigated and not others—that some knowledge is forbidden. A. V. Hill pointed out that this problem is not unique to science: "All knowledge, not only that of the natural world, can be used for evil as well as good; and in all ages there continue to be people who think that its fruit should be forbidden."

At stake is a principle most scientists hold very dear: freedom of inquiry. But that freedom is based on the premise that scientific inquiry is morally neutral, objective, and value free. Some, however, would argue that there are clear moral dimensions to the search for truth. There is obvious agreement that some *applications* of knowledge are wrong and must be banned, but this is technology and not fundamental science. But as ethicist Hans Jonas has noted, "not only have the boundaries between theory and practice become blurred, but . . . the two are now fused in the very heart of science itself, so that the ancient alibi of pure theory and with it the moral immunity it provided no longer hold."

Forbidden knowledge, new discoveries, dangerous technologies: Science clearly has the potential to challenge our values. We must make ethical decisions, but what guiding values will we employ? How will we balance the concern for truth and freedom of inquiry against concerns about human dignity and the ultimate consequences of biomedical techniques? This potential for good and evil intrinsic in our science and its applications should not surprise us if we have taken the biblical message seriously. Clearly, our exercise of dominion in the acquisition and application of biomedical knowledge is an awesome responsibility. The biomedical field has justifiably been referred to as revolutionary, primarily because of its challenge to our ethical and social values. This is an appropriate meeting ground for biology and Christianity.

## The Biomedical Revolution: A New Crisis

In his thoughtful book *Brave New People*, biologist D. Gareth Jones states the view that we are facing an unprecedented crisis:

> The current biomedical revolution, with its emphasis on genetic manipulation and genetic control, in vitro fertilization, quality control and high technology in medicine, may well have more widespread consequences for human life than either the Copernican or Darwinian revolutions—far-reaching as they were. We now have under our control the myriad of diseases that scourged us from the environment round about us, so that we can reflect on deficiencies in our own biological make-up. We can meditate on the quality of our genetic inheritance, and we are rapidly acquiring the ability to modify and redirect that inheritance.

From conception to the grave, and into the future, our knowledge is opening up possibilities that, in the view of many, belong to God alone. Procedures that were once barely conceivable are rapidly becoming realities.

### *The Scope of Biomedical Technology*

Much of modern medicine deals with the prevention and control of diseases that are well understood biologically, and the technology involved is often relatively inexpensive and noncontroversial—immunization, antibiotics, hormone therapy, and so forth. These technologies are usually spectacularly successful when their application is based on thorough understanding.

A second level of medical technology involves coping with diseases or conditions that are not well understood, or that are less easily prevented. The treatment involves compensating for the harmful effects of the problem—for example, bypass heart surgery, radiation therapy, aneurism repair. Because it involves a high degree of expertise, some very sophisticated equipment, and often hospitalization, this level of technology is often quite expensive. When it is controversial, the controversy over this level of medical technology usually surrounds questions of cost and accessibility rather than the appropriateness of the procedure. The "health care crisis" is generally understood as a crisis involving the high costs and low accessibility to medical care for many in society, at these two levels of technology.

However, a third cluster of technologies is highly controversial. Physician and ethicist Leon Kass has organized these more controversial aspects of biomedical technology into three groups, based on the major purpose of the technology: (1) control of life and death (2) control of human potentialities; and (3) control of human achievement. The first category, *control of life and death*, refers to those procedures that lead to greater control over fertility (and infertility) or that prolong life (or postpone death). Included in the former are techniques for preventing conception, for inducing abortion, and for circumventing infertility (for example, *in vitro* fertilization, artificial insemination). Among the latter, technologies that prolong life have become increasingly sophisticated, and include organ transplants, kidney dialysis, respirators, and artificial organs.

Biomedical technologies involving the *control of human potentialities* are often referred to as genetic engineering, with the distinction that humans, rather than other species, are the objects of the engineering. The intent of this kind of technology is to correct for harmful genetic traits, from treating genetic diseases to preventing them by genetic screening and counseling, and eventually to curing them by changing faulty genes. Perhaps the most disturbing current technology in this category is the use of amniocentesis to screen the fetus for a variety of genetic defects, a test that seems relatively harmless. However, the procedure presumes that defective fetuses will be judged unacceptable and are therefore to be aborted.

Finally, those technologies involving the *control of human achievement* include ways of altering people's attitudes, feelings and beliefs, and states of consciousness. Some of these technologies employ what is called biopsychological intervention—the use of mind-altering drugs or psychosurgery in order to bring about changes. Another class of technologies involve environmental manipulation—behavior conditioning, information control, and psychotherapy, for example.

Although this list of biomedical procedures is impressive, much of what is most controversial is also only a potential technology. There is still much that we cannot do—far more than what we can control at the present. But given some current trends, and our adoption of the biomedical model as a controlling paradigm, concern about the future is well grounded.

### *The Biomedical Model*

The fundamental paradigm that underlies modern medicine and its extension into health care is the *biomedical model.* According to this model, health constitutes the freedom from disease, pain, or defect; the normal human condition is health. The lack of health is considered to be a diseased condition, something abnormal; and this requires diagnosis and treatment to bring about the desired cure (and return to normality). According to the paradigm, illness conveys a special status; the sick person becomes a patient until cured, and is assumed to be both healthy and normal once the cure has been accomplished. Clearly, the model is strongly disease-oriented rather than health-oriented. There is a tendency to consider a human to be a machine that needs to be repaired if it isn't working properly.

Without question, this model has been highly successful in bringing "cures" for many diseased conditions. Critics, however, point to some significant problems with the model. It is impersonal; medicine becomes technology applied to the curing of a diseased state. The model tends to ignore the multidimensionality of health—the spiritual, emotional, social, and environmental contexts of the person. The biomedical model is also expensive; many of the medical technologies are of limited availability, and so both cost and availability tend to exclude many from its benefits. How do we keep health care from becoming a privilege only for those who can afford it? Is adequate health care a basic human right?

Another criticism of the model is that it generates unreal expectations of health. All diseases, pains, and defects are considered abnormal; the "normal" human life therefore excludes headaches, tensions, depression, deformity, and the like. Even death is seen as an abnormal event. One consequence of this tendency is what Gareth Jones calls the "overmedicalization of life," referring to the fact that we define everything from pregnancy to dying, from frustration to unhappiness, as a kind of illness. Then we proceed to apply biomedical technology in an often unsuccessful attempt to bring about a complete "cure."

The major fault of the model is that it equates medical care with health, as expressed by Aaron Wildavsky, dean of the graduate school of public policy at the University of California, Berkeley:

> According to the great equation, Medical Care equals Health. But the Great Equation is wrong. More available medical care does not equal better health. The best estimates are that the medical system (doctors, drugs, hospitals) offers about 10 percent of the usual indices for measuring health. . . . The remaining 90 percent are determined by factors over which doctors have little or no control, from individual life style (smoking, exercise, worry), to social conditions (income, eating habits, physiological inheritance), to the physical environment (air and water quality). Most of the bad things that happen to people are at present beyond the reach of medicine.

## Response to a Need: Bioethics

The Biomedical Revolution, based as it is on the successful application of the biomedical model, has brought both blessings and problems. Some things have become possible, or soon will be, that conflict with values held deeply by major segments of our society. Who is in control of biomedical technology? Modern society has become heavily dependent on this technology; but are we able to direct it in appropriate channels and see that it is not used in harmful applications? Can we even anticipate when harm will result from a given technology? How do we decide the complex biomedical issues that are already challenging our legal system?

Clearly, these questions will not go away. Such questions, and their resolution, have increasingly been addressed by a relatively young discipline, *bioethics.* Bioethics deals with the morality of human conduct in the area of the life sciences; it includes the medical sciences, but has applications outside of the medical field. By definition, a bioethical decision is an ethical decision that should be informed by biological knowledge. Before we turn our attention to several of the more difficult bioethical problems, we will briefly consider the characteristics of ethical systems.

### *Basic Components of an Ethical System*

In his excellent book *Ethics: Approaching Moral Decisions*, Arthur Holmes explains that ethics is about the *good*—those values and virtues that we should encourage—and about the *right*—our moral duties as we face practical problems. Recognition of the good and the right implies that ethics is a "normative" discipline—it tells us what we ought to do. Holmes goes on to describe the four ingredients to the development of ethical theory:

1. *Cases.* These refer to particular acts, and ask if an act is morally justified. Such decisions must be based on moral rules.
2. *Moral rules.* These are general guidelines that apply to various areas of life, for example, the rules that govern private property or the taking of a life.

3. *Moral principles*. These underlie our moral rules and are the broadest ethical concepts. They are considered to be valid in all cases; for example, the Christian principle of neighbor love.
4. *Bases*. Ethical principles are justified by reference to some philosophical or theological basis. This is the foundation for an ethical system.

Bioethics, then, is normative ethics applied to situations or issues that depend on some biological or physiological understanding. Some technologies or specific applications of biomedical technology are right, and some are wrong; at least, we might say, some are more moral than others. At this point, we have a fundamental problem: Our decisions must be based on moral principles and moral rules deriving from those principles, but which principles? Which ethical system shall we recognize? Christian ethics, we might be tempted to say; but a theory of ethics in our pluralistic society must be broadly acceptable. Arthur Holmes points out that Christian and non-Christian ethics differ mainly at the higher levels—the moral principles, and the bases of the system. Very often, the two are in agreement about the moral rules and the ethical cases. This is definitely not always true, however, and we need to examine more closely the role that might be played by Christian ethics in our society.

### *Christian Ethics and Morality*

Holmes speaks of the close link between religion and morality. The Christian religion, in Holmes's words, "identifies values to be propagated and virtues to be cultivated; and it . . . speaks to various kinds of behavior." Indeed, the Judaeo-Christian tradition is perhaps the major source of the Western world's moral beliefs. Holmes points out, however, that morality is not ethics. Morality refers to the involvement of right and wrong, good and evil in human behavior. Ethics implies a systematic study, and is normative. Morality thus is judged by the standards provided by an ethical system. In the case of Christian ethics, the Bible provides excellent material for constructing an ethical system.

The *theological base* for the system is simple: *to do God's will*. We are to love God, and love our neighbors as ourselves; we are to seek God's kingdom above all else. In addition, we find in the Bible the *moral principles* of *justice* and *love*, which are foundational to the *moral rules* that are also to be found throughout Scripture. It is then the Christian's responsibility to apply biblical morality to *specific cases*; where the Bible is silent, as in cases involving genetic engineering or population control, biblical principles must be applied. This constitutes *moral reasoning*, which is best done from within the context of an explicit ethical system.

However, what does this have to do with bioethical decisions? After all, God's law is not the law of the land, even though we might want it to be so. In this regard, it is very important that Christians work out of a framework of biblical morality. To the extent that we obey God's laws, we become the light that can shine in darkness and the salt that can preserve a society. Christians, therefore, have an obligation to develop a well-reasoned ethic and apply it consistently to their lives. This is an essential element in building and practicing a biblical worldview.

But what of society as a whole? Shall we who are Christians attempt to establish God's law as the law of the land? Shall we legislate morality? Arthur Holmes presents several positions on this latter question, and suggests that morals legislation is to be limited by the larger ethical consideration of individual liberty. And he points out that legislation is only one of many ways, and probably the most extreme, of influencing morality. Holmes concludes that legal moralism is both unjust and impractical in a pluralistic society, and that the mandate of government to legislate morality should be limited to matters pertaining to public justice and the social order. Admittedly, this is often hard for Christians to accept; indeed, we should be grieved when God's laws are not respected. On the other hand, the ethical principles that have emerged as foundational to bioethics are remarkably consistent with Christian ethics.

### *Principles in Bioethics*

The National Commission for the Protection of Human Subjects of Biomedical and Behavioral Research produced in 1978 a report based on four years of work with bioethical issues. In their report, the commissioners identified three ethical principles that seemed to be foundational to most bioethical problems. There was broad acceptance in our society for these principles:

1. *Respect for persons*. Individuals should be treated as independent beings, capable of making their own choices and judgments; also, where independence is diminished, as in the case of children or the aged, such persons are to be protected. Moral rules that derive from this principle include the sanctity of human life, truth telling, informed consent, confidentiality, the right to privacy, and the right to die with dignity.
2. *Beneficence* (active goodness or kindness). Embodied in this principle is the obligation to promote that which is good for people and, in doing so, to avoid doing harm. Further, it is important to remove harmful effects or conditions, and to prevent them from occurring if possible. The physician's Hippocratic Oath is closely identified with this principle.
3. *Justice*. The distributive principle that treats all persons equitably; all should receive what is fair or deserved. Injustice results when one is denied some benefit which they should receive, or when one is unfairly forced to bear some burden. The modern emphasis on individual rights is derived from this principle, as in the "right to health care."

Although these three principles are usually involved in bioethical decision making, the decisions that are made are not necessarily easy. Principles need to be interpreted. Sometimes the principles themselves come into conflict. Quite often the biological or medical information is incomplete or even faulty. For these reasons, the study of bioethics is often a study of difficult cases. Here all of the choices seem to be flawed and values and moral principles seem always to be in conflict, and the outcome is usually deciding which is the lesser of evils.

The case study approach can readily lead to the idea that bioethics is primarily a matter of deciding difficult cases. On the contrary, most bioethical decisions are part of the normal routine of scientific and medical practice and are noncontroversial. Physicians, nurses, and others involved in health care are usually clear on the ethical issues surrounding their work with patients and their families and, on the whole, the health care industry operates on the basis of high ethical standards. This day-to-day practice of ethics reflects most accurately the normative ethical status of our society, but unfortunately it receives far less attention than the spectacular cases brought to us by the media.

However, some decisions made by private individuals or by corporate bodies are subjected to challenge, and their resolution becomes a matter for our legal system to decide. When this happens, we as a society often find ourselves breaking new ground. Unfortunately, the sequence seems to proceed from knowledge to technology to application and only then, reluctantly and with much travail, to responsible decision making. Nevertheless, we do reach consensus on some matters; new bioethical decisions are being written into our laws. For example, the criterion of brain death as constituting legal death, once highly controversial, is now a matter of law in many states. Let us turn our attention now to current problems in bioethics.

## Exploring some Bioethical Problems

Two classes of problems confront us. First, we must ask if it is ethical even to gain fundamental knowledge if it has the potential for misuse or for inadvertent harm. For example, early in the recombinant DNA work, a group of the leading scientists in the field agreed to forego certain kinds of experiments until the dangers were adequately evaluated.

The other class of problems involves the application of knowledge already gained and translated into technology; is some particular application of the technology ethical? Is it, for example, ethical to fertilize human embryos *in vitro* (outside of the womb)? For both classes of problems, the ethical guidelines are seldom clear. The issues are usually complex, and frequently the problem is being encountered for the first time. And the biological component of bioethics reminds us that it is important to be thoroughly informed. We will briefly consider three of the areas that have created some of the most heated controversies: genetic defects, infertility, and abortion.

### *Genetic Concerns*

Mrs. Smith is forty-seven years old and pregnant. Because one of her sisters recently gave birth to a Down syndrome child, Mrs. Smith knows about the heartache of this condition. Because of her age, as well as the statistically significant chance that she is carrying such a child, the Smiths decide to find out. At eighteen weeks of her pregnancy, Mrs. Smith reports to the outpatient section of her hospital and undergoes amniocentesis, a procedure that samples the amniotic fluid surrounding her fetus. Cells sloughed off from the fetus are examined microscopically, and the pattern of three number 21 chromosomes is found, indicating positive for Down syndrome. The Smiths must now decide whether to abort the genetically defective fetus.

This is a hypothetical case, but it is representative of a class of ethical decisions being made daily in thousands of hospitals. Prenatal diagnosis has made it possible to detect practically all chromosomal disorders, as well as some seventy metabolic defects. Ordinarily, amniocentesis is only used when there has been some willingness to consider abortion if the fetus is defective; hence the two procedures must be linked in ethical concerns.

Screening for the presence of chromosomal or single-gene disorders may be performed before or after birth, and opens up a new set of ethical issues. There are advantages to such screening beyond the simple expedient of abortion. For some disorders, medical help can provide effective therapy when applied early. Also, persons who are known carriers for a defective gene are able to make informed choices about having children. Gareth Jones believes that there is a place for such screening, but that it must always be performed with the benefit of individuals or families in mind and never be allowed to lead to the prohibition of childbearing by individuals with known genetic defects. Jones sees a further danger in that genetic screening opens up the possibility of quality control.

> Quality control . . . recognizes the desirability of healthy individuals against the undesirability of unhealthy individuals. This can readily lead to the point of view that, given the technical means of producing healthy offspring, we are obliged to strive for such offspring, even if this is at the expense of unhealthy ones.

The ability to screen for genetic defects also introduces the real and potential uses of recombinant DNA techniques in human genetics (the ability to dissect out human genes and replace them or insert them into bacteria for the production of human proteins). This topic will be discussed in the next chapter.

### *Issues Surrounding Infertility*

In 1985, Mary Beth Whitehead signed a contract with William and Elizabeth Stern to be a “surrogate mother”—to bear a child for them that had been conceived through artificial insemination. The contract called for payment of all of the medical expenses plus $10,000. After giving birth to a baby girl (currently referred to as Baby M), Whitehead, who is the mother of two children, handed over the baby to the Sterns. She was so overcome with regret that she and her husband asked the Sterns for the baby's return three days later. Concerned about Whitehead's mental health ("We thought she was suicidal"), the Sterns reluctantly allowed the Whiteheads temporary custody of the child. Mrs. Whitehead did not accept her fee, nor did she sign over custody. Subsequently, the Sterns obtained a court order requiring that the infant be handed over to them; the Whiteheads fled with the child, but were tracked down and forced to relinquish custody. A superior court decision awarded custody of Baby M to the Sterns; the decision seems to have been based largely on considerations of the baby's welfare rather than on the rights of natural motherhood. An appeal led the New Jersey state Supreme Court to sustain the lower court decision on custody; paradoxically, the higher court ruled that such contracts for paid surrogacy amounted to illegal baby-selling, and declared the contract between the Sterns and Whitehead invalid. Whitehead was granted visiting rights, but the child will grow up in her father's home.

This case is illustrative of a cluster of issues that have emerged as a result of research and technology in human reproduction. Infertility affects one out of five couples in the United States. Some kinds of infertility may be treated with drug therapy, but in many cases more radical techniques are necessary. Artificial insemination (AI) is a relatively simple and safe procedure that is often used to overcome infertility. When it involves sperm donation from a donor (AID) instead of the husband (AIH), however, artificial insemination begins to open up some serious ethical issues. In particular, where it is used for surrogate motherhood, AID clearly raises concerns about the commercialization of reproduction and the legal questions of parenthood. An estimated dozen surrogate centers are presently in operation in the United States, conducting their arrangements in a current vacuum of legal regulation or restraint. Many states are now considering bills that would either ban or regulate surrogacy, a sign that the technology has finally gotten the attention it deserves. The Baby M case was initially considered a test of the legality of a contract versus the rights of motherhood. With the state Supreme Court decision, the case may well lead to a consensus that commercial motherhood is illegal, but that surrogacy might be allowed as long as the arrangement is based on good faith between the parties involved and the biological mother is allowed to change her mind after giving birth.

Another, even more controversial technology that has been developed as a response to infertility is *in vitro* fertilization (IVF). Louise Brown, born in England in 1978, was the first person to be successfully "conceived" in a test tube following removal of an egg from her mother's ovary. Some two days later, the early embryo was returned to her mother's uterus and proceeded to develop into a normal, healthy baby. Now that this procedure has been developed, it has opened up other possibilities: the freezing and/or donation of embryos and eggs, extensions of surrogate motherhood involving IVF, genetic manipulation and therapy, and even human cloning. To date, however, the procedure has been developed and used only to alleviate infertility.

Aside from future possibilities, IVF technology presents a number of serious issues: It is very expensive—who should have recourse to the procedure, and how much of our medical resources should be devoted to it? The procedure may lead to the production of "spare" embryos—are these unused embryos human beings? The spare embryos are often used for continued research in improving the technology—is this practice morally defensible?

The crucial issue raised by this technology—for the

present—is the status of the early human embryo. Do the spare embryos deserve the same respect as a fully developed human? If not, how much concern should we have for them? Gareth Jones, after discussing IVF in detail, concludes that there is no simple answer to the legitimacy of the procedure. He argues that the process is a legitimate response to infertility within marriage: "IVF is legitimate if it helps a married couple have a child of their own . . . an outcome of their own marriage . . . This is an acceptable goal, but it must be seen within the context of family love and the marital bond."

Jones contends that transfer of a human egg into the uterus of a woman from whom it did not originate is wrong, and also condemns some of the more futuristic applications of the technology such as cloning of humans and human-animal hybrids. Ambivalence, however, seems to be built into IVF. Are the spare embryos produced in the process of obtaining a successful implant significantly different from embryos lost during early pregnancy (miscarriages)? Clearly, many of these concerns are identical to the ethical issues surrounding abortion, to which we will now turn our attention.

### *Abortion: An Evangelical Watershed?*

Therapeutic abortion means abortion for medical reasons. Increasingly, the medical reasons for such abortions involve genetic defects in the fetus. However, the undisputed primary reason for most abortions is that a given pregnancy is unwanted. In 1985 (the last year for which complete statistics were available), 1.3 million abortions were reported for the United States, 81 percent of which were performed on unmarried women. The ratio of abortions to live births is currently 1:3. The public debate over abortion policy has been intense ever since the Supreme Court, in 1973, struck down both restrictive and moderate state abortion laws and determined that abortion in the first trimester of pregnancy was a matter for a woman to decide in consultation with her physician. The abrupt rise and continued high frequency of abortion testifies to a very strong pressure from pregnant women (and their partners) to seek to be released from the consequences of their pregnancy, namely, the responsibility of caring for a child. It also speaks of a profound failure of our society in the area of family planning and sex education—and, it is fair to add, morality.

The problem is particularly acute for teenage women (in many cases, still adolescents). More than a million teenagers become pregnant each year, four-fifths of them unmarried. Recent studies have shown that American teenagers are woefully uninformed about sex—in the technical sense. At the same time, they are literally deluged with sexuality by the popular media—TV, rock music, movies, and videos. One survey indicated that in one average year's viewing, some 9,000 episodes of suggested intercourse occured in prime-time TV shows. Social workers are unanimous in citing the strong influence of the media toward precocious sexuality. This information makes it quite clear that abortion can never be considered as an isolated phenomenon. The reasons that propel a woman (or an adolescent) towards an abortion are just as much of a moral concern as the abortion act itself; yet frequently our attention is narrowly focused on what happens in the abortion clinics alone.

The biological and ethical issues surrounding abortion are thoroughly discussed and thoughtfully presented in Gareth Jones's *Brave New People*. In his words:

> Abortion places upon the medical profession the mantle of both biological and social control, and this in turn presents many ordinary people with one of the most pressing and pervasive of human dilemmas. The planned destruction of human life brings us face to face with the meaning and finiteness of human existence. It forces us to examine the control we exert over future human lives and the reasons for bringing yet-to-be-born beings into existence.

In the first printing of his book (by InterVarsity Press), Jones carefully spelled out a position on abortion that included sanctioning therapeutic abortion under some circumstances (genetic defects like Tay-Sachs disease that involve serious mental and physical deformation). Because of this view, Jones was labeled as a proponent of abortion-on-demand by "certain sectors of evangelicalism within the United States." The outcome of their protests was finally a decision by InterVarsity Press to withdraw the book from the U.S. market. In a revised edition released by Eerdmans, Jones answers his critics and asks if indeed the abortion issue has become an "evangelical watershed":

> The heresy of which I appear to be guilty is that I cannot state categorically that human/personal life commences at day 1 of gestation. This, it seems, is being made a basic affirmation of evangelicalism, from which there can be no deviation. To adopt a position that deviates from the view that the embryo is anything less than a person demanding complete protection under every conceivable circumstance, is to exclude one automatically from the domain of evangelicalism.

He has a point. The explosive reaction to his book and its censorial removal by InterVarsity represents a symptom of weakness rather than the thoughtful Christian discussion and debate that this issue calls for.

To those who would like to work their way to a reasoned Christian response to this problem of abortion, in all of its biological, legal, and moral complexity, I can recommend (beside *Brave New People*) a book by philosopher Robert Wennberg, *Life in the Balance.* It was Gareth Jones's intention, as it is mine in this chapter, to encourage readers to think for themselves so that they can reach their own conclusions. This is true for all of the controversial issues raised by the biomedical revolution. It is encouraging that many of the Christian colleges have a course in bioethics and/or biotechnology. Here is the ideal context within which to develop a Christian ethic for living in a society that at times seems desperately in search of solid ethical ground, but stops short of a return to biblical standards.

## Conclusions

We began this chapter on the biomedical revolution with A. V. Hill's "ethical dilemma of science." Hill's point is valid—the accomplishments of our science can lead to both good and evil consequences. In our focus on biomedical science, we have pointed to the importance of bioethics as a frontier as well as a common ground for biology and philosophy. And we have looked at only a very few of the issues. This is an area of thought that calls for the involvement of committed and knowledgable Christians.

The effects of the Fall tell us that we will always misuse the possibilities opened up by biomedical technology. I believe that the benefits of that technology, however, are worth protecting and developing further. It is, in the best sense, subduing the earth to gain better control over our biological nature; but we must be careful that in the process we hold clearly in front of us the ethical and moral concerns that this work raises. An equally great problem is that of making certain that the benefits of our knowledge are brought to those who need them most, rather than simply to those who can afford them. This too is a Christian imperative, one that is sometimes lost in the complexity of the issues.

### Summary

Biomedical knowledge and its technological applications represent an important exercise of human dominion. But the field of biomedicine has also brought consequences and possibilities that are clearly harmful. This situation is illustrative of A. V. Hill's "ethical dilemma of science," where scientific knowledge can lead to both good or harmful consequences. In its remarkable new capabilities and its capacity to challenge our ethical values, the biomedical field clearly deserves being called revolutionary.

Much biomedical technology is routine, successful, and noncontroversial. However, there exists a cluster of technologies that are highly controversial, including procedures that are aimed at controlling fertility, prolonging life, diagnosing genetic defects, and altering human behavior. Although many concerns are based on technologies that have not yet been developed, our acceptance of the biomedical model as a controlling paradigm for medicine justifies giving attention to these future possibilities. This approach assumes that health is the freedom from disease, pain, or defects, and therefore the delivery of medical care is equated with health. The model generates unreal expectations of health, and is highly expensive and often impersonal. Yet we have become highly dependent on biomedical technology. How should it be controlled?

Bioethics represents a response to the need to examine the morality of human conduct in the life sciences, especially the medical field. (For those unfamiliar with philosophy, the basic components of an ethical system are outlined). The question arises, however, of which ethics to adopt in a pluralistic society. It is the responsibility of Christians to develop a biblical ethic and apply it consistently; this is vital to the development of a Christian worldview. Although we might like to see God's law become the law of the land, such legal moralism is both unjust and impractical in our society as it exists. Remarkably, the foundational principles of bioethics are consistent with Christian ethics (but not inclusive).

Three ethical principles are broadly accepted and applied in biomedical situations: *respect for persons*, *beneficence*, and *justice*. The high visibility of difficult cases has led to the misconception that bioethics basically means solving these cases. On the contrary, most bioethical decisions are carried out daily, and with high ethical standards, by health care practitioners. Yet some knowledge does become used in technology before the ethical problems are resolved.

For example, genetic screening of the fetus has led to difficult decisions about whether to abort a fetus with a genetic disorder. This in turn raises questions about quality control of humans and the prospect of human genetic engineering. Further, the strong desire to overcome infertility has led to a number of disturbing possibilities, including surrogate motherhood and *in vitro* fertilization (test-tube conception). Abortion is a particularly difficult issue involving social, religious, and moral concerns. Censure of a book authored by Christian biologist Gareth Jones has pointed to the dangers of allowing a controversial and complex ethical issue to become a basis for judging Christian commitment.

As fallen people, we will always misuse the possibilities opened up by biomedical technology, but the benefits far outweigh this danger. It is highly important, however, to bring out the ethical and moral concerns that this work raises. It is also important to make certain that the benefits of this technology are extended to those who need it most.

*Chapter 11*

# THE GENETIC REVOLUTION

There has hardly been a more decisive breakthrough in the whole history of biology than the discovery of the double helix.

ERNST MAYR

## DNA, the Essence of Life

Children usually resemble their parents but are never exact copies of them or of each other (except for identical twins). These two tendencies—transmission of characteristics to offspring, and variability within a species—are signals of a profoundly important fact: Living organisms are the product of a program, and the program is subject to change.

We are now certain that the program is written in specific sequences of four kinds of subunits (*nucleotides*) linked together in long molecular strings that we call DNA (*deoxyribonucleic acid*). The sequences represent information that directs the growth and activities of the cells, tissues, organs, and behaviors that make up a living organism. The directing information also controls the manner in which cells may respond to or translate the information, and the manner in which the information is packaged to be passed on to the next generation. It is a complex program, but we are learning how to read—and change—its messages.

The story of how we have come to this knowledge begins in earnest in a central European monastery garden in the mid-1800s, with the Abbot Gregor Johann Mendel. Every biology text describes Mendel's careful work with the common garden pea, the publication in 1866 of his results, the almost total neglect that greeted his work, and its subsequent "discovery" in 1900. Following that, the science of what we now call classical genetics entered its golden age. Looking back, we can identify Mendel's most important discovery: the fact that inheritance is a matter of particles (*genes*) that occur in sets, one of which comes from each parent. Geneticists quickly discovered the *chromosomes* as those structures in the cell which contained the genes, and heredity was understood as the product of genes on chromosomes. This was important knowledge, with many immediate applications. But there was much more to come.

A separate history traces the recognition of DNA as the chemical basis of heredity and the discovery of the double helix—again, the stuff of every biology text. The day that James Watson and Francis Crick realized that they had worked out the double helical structure of DNA, Crick boldly announced in a pub across the street from Cambridge University's Cavendish Laboratory that they had just discovered the secret of life! In essence, Watson and Crick had gathered information from numerous other workers to construct a model of the chemical structure of DNA. Their model demonstrated that DNA exists as two long chains bonded to each other by weak bonds (hydrogen bonds) between the nucleotide bases arranged in a sequential fashion along the chains. Although the four bases (adenine, cytosine, guanine, and thymine) occur in a variable sequence along a given chain, each one specifically pairs with its opposite base on the other chain: adenine always with thymine, and cytosine always with guanine.

Watson and Crick published their findings in 1953, concluding with a classic understatement: "It has not escaped our notice that the specific pairing we have postulated immediately suggests a possible copying mechanism for the genetic material." At once it was apparent that the sequence of bases carried the information content, the program; and that within the cell the information could be copied exactly if the two strands separated and directed the synthesis of complementary DNA strands. The genetic variability that exists between different members of the same species was now understood as the result of small differences in the base sequences that make up the genes.

### *Molecular Biology*

The discovery of the double helix opened up an exciting new field—*molecular biology*. It has added so much to our knowledge of heredity and cellular function that the term "revolutionary" sometimes seems inadequate. We have moved from understanding how information might be encoded (Watson and Crick's work), to a precise deciphering of the code itself; and we are now at the threshold of being able to read the entire genetic instructions of a living organism. In fact, if viruses can be considered organisms, this has already been accomplished; we know the entire base sequence of numerous viruses.

However, the base sequences are only the very beginning of understanding how heredity is controlled. These sequences—the encoded information—exist in the context of subcellular structure, cells, and tissues, all of which exert a crucial influence on how that information is read and what happens to the products once the information is translated into proteins. The situation might be compared with a book on solid state physics that deals with the design and manufacture of semiconductors—written in Russian. A non-Russian reader can identify the letters, can describe the sequence in which they are found, and in time may learn to translate the book into a familiar language. The knowledge represented by the book is only gradually acquired, and even after translation it may be so complex that the reader is unable to comprehend it. In terms of the genetics of living organisms, we are fairly competent in reading the sequence, a process requiring sophisticated knowledge. We are not very good yet in translating the information, and we are still a long way from complete comprehension. Nevertheless, the knowledge already gained has given us new powers and opened up previously unimagined possibilities.

### *The Human Genome*

Molecular biologists have raised the prospect of sequencing the entire human genome (genetic program). This means determining the sequence of the approximately 3 billion nucleotides that are found on the human chromosomes. This is an enormous and costly task, but it is judged to be technically feasible. Currently, sequencing of human genes is performed in many laboratories and for different reasons; the genome project would provide direction and the necessary funding, and at the same time eliminate inefficient competition between laboratories. In the words of Walter Gilbert, a leading Harvard biologist favoring the project, "The total human sequence is the grail of human genetics."

A less expensive and more immediately useful goal is to establish a complete "map" of the human genome, which means to determine the physical location of the genes on the chromosomes. This information would be greatly helpful in gaining insights into the well-known genetic defects as well as genetically complex diseases such as heart disease. This is viewed as phase one in the proposal; phase two is learning the sequences, and phase three, in Gilbert's words, is "the complete understanding of all human genes." Because the mapping and sequence information is clearly capable of being put to commercial use, Gilbert believes that it should be possible to patent the gene sequences once they are learned—a proposition that has been challenged by others in the field.

The human genome project became a reality in 1988, when a new Office for Human Genome Research was created at the National Institutes of Health. It will be headed by James Watson, an indication of the strategic significance of the project within the scientific community. Funding is expected to reach $200 million per year by 1992, and the ultimate cost of the effort is projected at $3 billion.

### *Recombinant DNA*

Learning the sequence of the human genome is not the only controversial enterprise of the genetic revolution. Clearly, knowing the sequence in an organism's genes opens up the possibility of changing it—manipulating the genetic material and thereby changing the outcome as it dictates the growth of an organism. This is no bizarre futuristic prospect; genetic engineering is already with us in some forms. Twenty years after the double helix was announced, a new technology appeared as a result of work on bacteriophages, the viruses that infect bacteria. We are now living in the age of *recombinant DNA* (*rec*DNA). This is the technology known as "gene splicing"—the process whereby pieces of DNA from different sources are joined together. Gene splicing had its inception with the discovery in 1973 of *restriction enzymes* in bacteria. These are enzymes that respond to a viral attack by cutting the invading viral nucleic acid into pieces which can then be digested. A large variety of bacterial restriction enzymes have been discovered and purified, each one cutting the nucleic acids at different, known combinations of base sequences. It is, in fact, the use of these enzymes that has made sequencing the human genome technically possible.

With the aid of other enzymes and *plasmids* (particles of circular DNA found in bacteria), it is possible to cut DNA—from any source—and attach it to a plasmid, which is then introduced into a new cell (usually a bacterium). If the appropriate control information—also DNA—has been included, the new DNA (recDNA now) carries out its normal function in the new cell. Thus bacteria can be engineered to manufacture human proteins. Indeed, the possibilities for manipulating genetic information opened up by this technology are practically limitless. In his thoughtful book about the implications of this new technology, *The Double-Edged Helix*, biochemist Liebe Cavalieri comments on this potential:

> Another paradigm had been created; new biological domains, not heretofore readily accessible, were open to investigation. For example, the study of the structure and function of mammalian genes was greatly facilitated, for functional genetic subunits could be separated, reassorted, and studied individually. The synthesis of drugs and foreign proteins in bacteria was made possible; genetic engineering of higher organisms by the use of recDNA techniques became an enticing prospect for some scientists. The overwhelming potential of recombinant DNA technology was apparent from the outset. The excitement spread like wildfire throughout the science community, rekindling dying embers in some laboratories and fanning flames in others, both large and small. And the fires still burn.

RecDNA methods are leading to some remarkable achievements. For example, because of the capabilities of recDNA to detect DNA mutations, diagnosis of numerous inherited and acquired diseases has been expanded and made much more precise. New insights into the mechanisms of enzyme action are being gained because of the possibility of making precise changes in the genes that control the synthesis of enzymes. Genetic control of the process of embryonic development is now being worked out with the use of recDNA methodology. The transfer of desirable genes into various kinds of plants, including the cereal grains, is now being accomplished. Genes regulating the production of blood cells in humans have now been cloned and promise to lead soon to practical use in treating many different blood deficiencies. Experimental work suggests that it soon will be possible to treat some genetic diseases by transplanting bone marrow cells into which the normal gene has been inserted.

Certainly one of the most crucial current applications of recDNA methodology has been its thorough involvement in research dimensions of the AIDS (acquired immune deficiency syndrome) epidemic. The AIDS virus acts on cells of the immune system by inserting its own genes into the host's chromosomes, which is in fact a form of genetic engineering. The discovery of how the AIDS virus acts on cells, the details of its viral activities, the genetic information present on the viral nucleic acid, and the potential strategies for therapy against the virus have depended intimately on techniques made possible through recDNA research.

What is particularly revolutionary about recDNA is the fundamental change it has wrought in the research program, the paradigm, of molecular biology. Before recDNA, the task of molecular genetics was to understand the structural and functional aspects of heredity at the molecular level. This was basic, or "pure" science. Now it is possible to manipulate the genetic material in ways never before imagined. This new power comes with a remarkably low price tag; many of the techniques can be easily accomplished in a small laboratory with a limited budget. Predictably, these new developments in molcular genetics have stimulated both serious concern and great optimism. But the genetic revolution is in full progress. We are in the midst of an exponential increase of knowledge of the genetic programs of living organisms, and we are fully involved in putting that knowledge to use.

## Applications and Expectations: Biotechnology

The prospect of engineering organisms using the new recDNA technology was almost immediately translated into a new commercial venture: *biotechnology*. Biotech companies were formed in the 1970s around many of the prominent scientists in the field, in great anticipation of lucrative applications of the new procedures. The first venture was in the production of biological compounds for medical use; human insulin, growth hormone, and interferon are three of the compounds now being produced by engineered bacteria.

Because of the concentration of molecular genetics research centers in the Boston area, Boston has become a world center for the biotechnology business. Some forty biotech companies in and around the city compete intensely for the development of new drugs like tissue-type plasminogen activator (tPA), which dissolves blood clots that cause heart attacks.

Other applications currently in the research and development phase involve the intentional release of genetically engineered organisms into the environment. These include the use of genetically engineered plants and microorganisms for agriculture (for example, microorganisms that would protect plants from freezing); and applications in the area of mining, waste treatment, and the cleanup of chemical spills (for example, microbes that would improve the recovery of low-grade mineral deposits).

### *Hazards and Risks*

It was immediately recognized that the new technology might also involve new risks. New combinations of genetic information inserted into microorganisms caused concern about what might happen if any of these engineered forms accidentally escaped from their laboratory setting. Visions of recombinant bacteria or viruses creating epidemics in humans, or doing unexpected things in the environment, led to the Asilomar Conference in February 1975. At this conference, leading figures in the molecular genetics field argued in favor of a moratorium on certain kinds of recDNA research until proper safeguards could be established. The final report from the conference included recommendations for laboratory safety standards for the various kinds of research. These recommendations eventually led to the Recombinant DNA Advisory Committee (RAC) at the National Institutes of Health (NIH) and the promotion of guidelines that were made mandatory for all research sponsored by NIH.

In the decade of research since Asilomar, the technology

has changed radically—in the words of Nobel laureate David Baltimore, "from rudimentary but promising to sophisticated and routine. Biologists today do experiments inconceivable in 1975." The laboratory safety question has all but disappeared; no known death, injury, or environmental contamination has been traced to recDNA research. The record has been so good that the RAC guidelines were relaxed in 1982; and after considering the risks and the record, Congress decided not to legislate controls.

However, proposals for the intentional release of bioengineered organisms into the environment have raised new concerns. What will be the ecological impact of these new forms of life? Can we even imagine all of the potential dangers of their release? History informs us that the introduction of new microorganisms into unfamiliar environments can lead to great harm—Dutch elm disease, chestnut blight, and the myxoma virus to control rabbits are all examples. The task of assessing the ecological risks involved with intentional release falls on the Environmental Protection Agency (EPA). The consensus of a recent seminar on biotechnology and the environment indicated that EPA has a difficult task ahead. In the view of many at the seminar, the technology and its impending applications have expanded far more rapidly than our ability to assess and manage the risks.

The National Academy of Sciences (NAS), responding to what it saw as undue concern about introducing genetically engineered organisms into the environment, stated in a pamphlet published in late 1987 that "adequate scientific knowledge exists to guide the safe and prudent use of recombinant DNA-engineered organisms in the environment." The pamphlet was immediately challenged by leading ecologists, who called the report unbalanced and inaccurate. They cited the lack of ecological input, and pointed to the strong bias toward genetic engineering in the makeup of the panel that created the report. It is clear that the NAS publication was more of a political than a scientific document, and that it probably failed to accomplish its basic purpose of reassuring the public. This is clearly an unresolved issue.

### *Concerns About Human Genetic Engineering*

Quite a different set of concerns surrounds the application of recDNA to humans. The most obvious application is in the correction of genetic defects—*gene therapy*. This can be accomplished by one of two methods: (1) *somatic gene therapy*, inserting a gene into somatic tissues to cure or prevent development of the defective condition; or (2) *germ-line therapy*, inserting the normal gene into the germ line to prevent the defective gene from affecting the next generation. The first method is technically much more feasible than the second, and has recently been accomplished for several defective conditions in laboratory mice. In such genetic disorders, the defect is identifiable; the cause of the defect is a single defective gene, and the therapy involves insertion of a normal gene in tissues where the gene's effects will be "expressed," or produced. An example of a potentially correctable defect is the Lesch-Nyhan syndrome, a disorder leading to severe mental retardation and self-mutilation.

In anticipation of the application to humans, a subgroup of RAC has now established guidelines that will govern the human gene therapy experiments expected to occur soon. Among these guidelines is the call for public review of such proposals, and the refusal to consider any experiments involving germ-line therapy. The committee backed away from expecting the researchers to address the complex social and ethical concerns raised by their work, noting that such concerns are currently not well addressed by ethicists. Indeed, somatic gene therapy seems to present no ethical problems that go beyond normal biomedical ethics, unless one assumes that it will open the door for germ-line therapy.

The introduction of genes into the human germ line is a different matter, both technically and ethically. The technology is presently not competent to accomplish this, and may not be so for some time; claims that scientists might begin "playing God" by improving the human species seem to be highly exaggerated. However, the possibility of applying recDNA technology to the human germ line has alarmed many observers, including Jeremy Rifkin, a self-styled activist in environmental and scientific issues. Rifkin's concern is that the new biotechnology will give scientists the capability to alter the genetic future of the human race, perhaps to "improve" it, which he believes should not be attempted. Rifkin is also firmly opposed to the development of recombinant hybrid organisms, on the grounds that such work transgresses natural boundaries and in fact only benefits the biotech companies and not society. In his book *Algeny* (meaning humanity's use of biotechnology to change living things) Rifkin suggests that this will lead us into the next great (but dark) epoch of history.

> Today, our biotechnical arts merely imitate nature. Tomorrow, they will subsume it. Our children will be convinced that their creations are of a far superior nature to those from which they were copied. They will be the algenists. . . . They will redefine living things as temporal programs that can be edited, revised, and reprogrammed into an infinite number of novel combinations. The algenists will change the essence of living things. They will upgrade the performance of living systems. They will program entirely new biological processes. They will seek to transform the living world into a golden treasure trove, perfectly engineered, optimally efficient state.

Is Rifkin simply an alarmist clamoring for attention, or is he doing us a major service by drawing our attention to issues that vitally concern us? Is biotechnology really a serious threat to our future, or just another technology in need of more careful scrutiny and regulation?

### *Science, Technology, and Society*

Other voices, not as strident or controversial as Rifkin's, speak of recombinant DNA technology as something less than pure blessing. In *The Double-Edged Helix,* Cavalieri expresses a fundamental concern that recDNA research has become the handmaiden of industry rather than a tool for gaining scientific knowledge about how genes work. He cites the direct and immediate connection between recDNA work and biotechnology, and fears that the pursuit of commercial profit has become the primary driving force behind most recDNA research. Given this trend, the scientist-technologists cannot be trusted to guard the public interest. Like all technologists, they will first serve the industrial structure that sustains them, and only secondarily respond to human needs. And they will likely resist efforts to regulate the technology, claiming that recDNA research is our best hope for understanding cancer and heart disease, that regulation interferes with the freedom of inquiry that scientists should enjoy.

Cavalieri calls for more control over biotechnology, and for a more responsible practice of science, where scientists develop and apply a social conscience to their work and the technology it spawns: "The scientist who insists on placing the responsibility with the technologists is rather like a person who makes a useful box of matches and leaves it in a room full of pyromaniacs." The freedom of inquiry should not extend to biotechnology; there are grave risks associated with it that must be evaluated. Science has a higher calling—to be the servant of more fundamental human values than a hypothetical "freedom of inquiry."

Cavalieri's main concerns are the essential irreversibility of the creation and release of bioengineered organisms—the envionmental problems cited earlier—and the strong and well-entrenched tendency of technology to proceed with its agenda. Any risks to the public should be balanced by real benefits, and it is questionable if the benefits are at all in the public interest. Cavalieri calls the scientist to:

consider the real significance of science—his work—in the world picture. Where is the present direction of research likely to lead? How will the results ultimately be applied? What impact would they have on the quality of life, judged in the overall context. . . . Conscience literally means "with knowledge," and surely that is how every scientist would wish to proceed.

As a scientist who helped to create the field of molecular biology, Cavalieri's call deserves the thoughtful consideration of all who enter the sciences.

**Ethical Issues**

Jeremy Rifkin's Foundation on Economic Trends issued in June of 1983 a resolution on the theological issues in genetic engineering. Sixty-three religious leaders and scientists across the spectrum of political and religious thought signed the resolution. The resolution called for a complete ban on "efforts to engineer specific genetic traits into the germ line of the human species." Rifkin uses the slippery-slope argument: Once we embark on this line of research and technology, there is no stopping point. The perfectability of the human species is the logical goal, and great evil could be done in the attempt to reach that goal.

It is only fair to point out that the concerns of Rifkin and Cavalieri (and many others) are balanced by a substantial response by the scientific community that considers the dangers and the ethical ramifications of recDNA technology. For example, EPA is proceeding on the assumption it has jurisdiction over any intentional release of bioengineered organisms into the environment, and is in the process of developing regulations under the authority of the Toxic Substances Control Act. RAC has recently drafted a document, "Points to Consider in the Design and Submission of Human Somatic-Cell Gene Therapy Protocols," which represents a distillation of fifteen years of ethical discussion. This document has been published in the *U.S. Federal Register*, and has been supplemented by the creation of the RAC review committee referred to earlier. Recall that the committee has refused to consider any germ-line therapeutic proposals.

The difficulty of developing regulations and sorting out all of the ethical concerns is apparent. We are entering uncharted waters, and are trying to determine the potential harm versus the good that might be done by technologies that in many cases have not yet been developed. Much of Rifkin's concern—remaking and perfecting human beings—is presently outside the realm of possibility. However, there is value in raising such concerns, in that it forces us to consider the possible relationship between steps taken now and future potentialities. We are forced to open questions for discussion that might ordinarily be dismissed as unnecessary intrusions into scientific matters. There should be no reluctance to discuss these matters publicly. In this light, the pamphlet put out by the National Academy of Sciences seems to be an attempt to stifle full consideration of some of the dangers of biotechnology. It should be possible to speak to the futuristic concerns raised by Rifkin, to agree that such goals as engineering people and attempting to redesign the living world are ethically and philosophically unacceptable, even though we suspect that they are largely technically unattainable. The road to such a consensus, however, is not at all well defined.

### *Christian Responses to RecDNA*

It is entirely appropriate to consider the information encoded in the nucleic acids of living organisms as God's instructions to living things, as suggested by biologist David Wilcox:

> God calls all things into existence and holds them in existence by the Word of his power. . . . Assuming that Christ (the Word) indeed so orders the creation, that divine organizing command might be considered analogous to the role of DNA in living organisms. For living things to obey the "biotic command" of God, they must "obey"

the information in their DNA, organizing their structure by that inherited pattern. In a sense, therefore, the information of the DNA might be considered a small encoded portion of the commands of God . . .

DNA, then, is the medium that carries God's commands to living organisms, and they "obey" his commands as their lawful response. This is a profoundly important insight into the covenant between God and his creation. He commands, and the creation obeys. It follows, then, that we are in the scientific enterprise of the genetic revolution, reading God's commands to living things at their most fundamental level.

However, having learned to read God's instructions, we are also revising them, and this is a powerfully significant step. How should we view this? It seems clear that recombinant DNA research and the genetic engineering that it leads to provide us with another example of A. V. Hill's *ethical dilemma of science*. Both good and evil can proceed from this knowledge; shall we stop its development because of the evil that could occur? Some would argue that we should. Here are the views of Henry Stob, in his book *Ethical Reflections:*

> A conviction governing a great deal of the scientific concern with man is that man is not a finished product of creation but is an unfinished, malleable and open-ended something, which, having been produced by mother nature, is being moved by evolutionary forces into a promising future. It is this conviction which justifies for many scientists the various forms of genetic engineering. Bio-medical science, in this context, is not concerned, as it was in the past, simply to support or heal; it is concerned to program and direct, and in this way to be as creative as nature itself. . . . Because man is divinely structured I find it hazardous, if not impious, to tamper with the genetic core. To tamper with genes seems to me to outrun God into an unknown future and to exercise an elective discrimination mere men do not possess.

Yet it would seem that the current goals—if we set aside for the moment the profit motive—are for the most part consistent with the Genesis mandate to subdue the earth and have dominion over it. We as stewards have the responsibility to understand the creation and to create a culture; both of these tasks are represented in recDNA research. The work has a great potential for uncovering the patterns of creation, and therefore can demonstrate God's glory and result in his praise. And we are to pursue justice and do good works, both of which are serious moral imperatives for the medical applications of recDNA research and technology. Should we not welcome the technology that produces large quantities of human growth hormone, or cures genetic diseases, or has the potential to bring AIDS under control?

But we are also stewards over all creation, and should be bringing healing to it rather than harm; therefore the environmental risks of recDNA become important. And we should suspect the technological connections of recDNA work, because they are undoubtedly tainted with greed. Because we know what manner of creature we are, we should insist on continual surveillance of recDNA research and thorough regulation of genetic engineering technology. Christians in the life sciences should be interpreting this work for the church and making clear to society what the moral choices are.

These are not simple matters. As it unwraps the secrets of living things, recDNA research will undoubtedly do more to multiply human arrogance than to bring praise to God. Is this not what lies at the root of the notion that we can patent living organisms if we have altered them genetically? It is entirely possible to develop biotechnologies that will bring help to those who need it most—the poor, those in Third World countries who suffer from hunger, parasites, and diseases. Is there any hope that these will be the priorities of a profit-motivated biotech industry? Some day, perhaps already, genetically altered organisms will be released to the environment and will join the creation. Is it not morally wrong to do this when we are not at all certain of the outcome? And then there is humankind, and the potential to change our own genetic makeup. Is it right or is it wrong to develop somatic cell gene therapy when we suspect that the next step might be germ cell therapy?

RecDNA research and the biotechnology it is spawning are clearly here to stay. Some would favor a complete moratorium on such activities, and I respect that viewpoint. It reflects a deep concern about the potential for this work to do great harm to the integrity of the creation, whether it is to ourselves or to the other creatures and the environment we share together. Others see the great potential of this work to unlock many of the secrets of life, and therefore to bring into our knowledge more of what God has done. Given this view, it is clearly vital to make sure that this knowledge is used in a context of responsible stewardship, and used with great respect and awe for the remarkable work of creation that it is.

## SUMMARY

Discovery of the mechanisms of heredity began with Mendel's work and led to the recognition of DNA as the hereditary material. When the structure of DNA was worked out by Watson and Crick, the field of molecular biology began in earnest. The genetic code has now been deciphered, and it is possible to read the genetic information contained on the chromosomes of living organisms. However, that information only functions in the context of the cells and tissues of organisms, and we have much to learn about how the genetic information directs the growth and activities of a living organism. Among the possibilities opened up by these discoveries is mapping and sequencing the human genome, and there is now a national commitment to this project.

It is now possible to manipulate the genetic instructions of an organism using the techniques of recombinant DNA research. DNA from any source can be inserted into a different organism—usually a bacterium—and will then carry out its normal function in a different organism. This gene splicing has led to some remarkable achievements, enhancing our understanding of many life functions and leading to new diagnosis and treatment of disease states. In particular, the new technology is helping in research on the AIDS virus. There is intense competition now to develop commercial products using recDNA technology—called biotechnology.

Early concerns about the hazards of this work led to the promotion of clear guidelines for laboratory safety, and in the ensuing years no known death, injury, or contamination has resulted from recDNA work. However, proposals to release bioengineered organisms to the environment have raised serious concerns about the unknown ecological impacts such organisms might have.

Proposals to apply recDNA methods to humans involve attempts to correct genetic defects—gene therapy. Somatic gene therapy, the less controversial of these applications, seems to raise no new ethical problems. Germ-line therapy, although technically not yet feasible, appears to open up the prospect of changing the genetic future of humanity, and is viewed with alarm by many. Indeed, the social and ethical implications of recDNA technology are being debated broadly. As yet there is no clear consensus on limits to be placed on this work.

A Christian approach to recDNA begins with recognition that the information encoded in living organisms is part of God's instructions to living things. Organisms obey these commands as their lawful response to the covenant between God and his creation. We are not only reading God's commands, however, but we are also undertaking to revise them for our own purposes. Is this legitimate? The answer depends on the use to which this knowledge is put, for it can produce both good and evil. It is a great stewardship challenge to guard the integrity of the creation and yet to use this technology for accomplishing good ends.

*Chapter 12*

# THE ENVIRONMENTAL REVOLUTION

> The French use a riddle to teach schoolchildren the nature of exponential growth. A lily pond, so the riddle goes, contains a single leaf. Each day the number of leaves doubles—two leaves the second day, four the third, eight the fourth, and so on. "If the pond is full on the thirtieth day," the question goes, "at what point is it half full?" Answer: "On the twenty-ninth day"
>
> LESTER R. BROWN

## Environmental Crisis: Seeds of a Revolution

Lester Brown is president of Worldwatch Institute, a Washington-based organization that investigates global environmental issues and publishes its findings in an annual "State of the World" report as well as in numerous "Worldwatch Papers." Brown's thesis in *The Twenty-ninth Day* is that our global lily pond—the environment that is our life-support system and the source of all of our resources—may already be half full. Given the nature of exponential growth, in one more generation it will be completely full; we will have completed the Genesis mandate to "fill the earth." If this is true, our civilization will either have to learn how to live within the limits imposed by the environment, or else it will have to cope with the consequences of large-scale disintegration of the biological systems that sustain life—consequences that could include famine, resource scarcities, and a major deterioration of public health on a global scale. This, in a nutshell (or a lily pond), is the *environmental crisis*.

The environmental crisis is no ordinary crisis. It has been

coming on very gradually, but it could rather suddenly intensify, depending on how we respond. It is a cluster of issues, all related in a complex web, many of which have only been considered serious for a decade or two. There is no easy solution to this crisis, no new technology or political decision that will make it go away. I have referred to the "environmental revolution" in the chapter title, yet it should be evident that no such revolution has occurred. The Darwinian, biomedical, and genetic revolutions that we considered in previous chapters represent major changes in human thought and human affairs as a result of new knowledge in the life sciences. The environmental revolution, as I intend the meaning, is a revolution of the future. It will come about as humankind is increasingly forced to live within the sustainable limits of the global environment. My thesis is that by the time one more generation of people is born—thirty years or so—our way of life will have been changed so radically that, looking back, we will recognize that a revolution has taken place. There are major changes coming.

### *The Future: Doom or Utopia?*

Future-gazing is a time-honored preoccupation, quite often merely a playground for the imagination. A new breed of futurists has appeared in recent years, and they have a serious agenda: they are looking into the future in order to influence the present. They tend to divide naturally into two opposing camps—the *Cornucopians* and the *Jeremiads*. When Cornucopians gaze into their crystal ball, they see a bright future. They are not disturbed by current trends (things have never been better!) and reassure us that our problems are not all that serious—the problems have been exaggerated out of proportion. Where there are problems, there are also solutions; Human ingenuity and new technology are rich resources that are quite capable of solving the future problems that the Jeremiads are worried about. Cornucopians argue that the gloom and guilt that often accompany the pronouncements of the Jeremiads are counterproductive, and can often lead to despair. Their message, on the contrary, is designed to encourage and affirm people. They often come from the ranks of economists, businesspeople, and politicians.

Just as Jeremiah announced dire consequences unless the people of Judah repented (and got himself thrown in prison for his message), the Jeremiads see a gloomy future and warn us to change our ways. Their strategy is to take current trends and project them into the future, examining in some detail what kind of future we will have if things don't change. Their message is that things are not going well—the environmental crisis is a reality—and we need to do something now if we want to prevent future disaster. This point of view is common among environmentalists, many resource specialists, and demographers (those who study the human population).

The debate between these two camps comes with strong political overtones. The Cornucopians would endorse the old adage, "If something isn't broken, don't fix it!" We're doing fine, and would be unwise to interfere with our social, political, and economic systems. The Jeremiads argue that major changes are essential, usually calling on the government to bring them about. Not surprisingly, the debate often divides along conservative/liberal lines.

Regardless of political persuasion, be assured that our choice is not simply doom or utopia, radical action or complacency. These are not viable Christian options. Specifically, what are these two camps arguing about?

### *Major Components of the Environmental Crisis*

Environmental scientists are in broad agreement about the primacy of the following issues:

1. *Pollution*—the degradation of air, land, and water as a result of the release of chemical and biological wastes.

2. *Population*—related to hunger and poverty in the less-developed countries, and related to resource consumption and pollution in the industrial countries.
3. *Resources*—food supply, the approaching limits on nonrenewable mineral and energy resources, and the deterioration of renewable biological systems that provide food and fiber.
4. *Biological depletion*—the destruction of tropical forests, the extinction of species, and the relentless removal of natural habitats by human activites.

The issues are interrelated in many ways, and each issue is in reality a complex problem. An analysis of any one issue immediately uncovers the involvement of economic concerns, technologies, political systems, and ethical norms—fundamental elements of all human societies. We will shortly explore these four major issues in more detail.

### *The Question Before Us: The Global Possible*

In May 1984, seventy-five leaders of scientific, governmental, industrial, and citizens' groups from twenty nations met to address the following question: "*Can the world reverse current resource and environmental deterioration while at the same time promoting a better quality of life for all and achieving a marked improvement in the living standards of the world's disadvantaged?*" The conference, entitled "The Global Possible: Resources, Development and the New Century," was sponsored by the Washington-based World Resources Institute. There was overwhelming agreement about the need for concerted action. Projections of current trends into the twenty-first century indicating a bleak future were accepted by the conference members. But they were also agreed on the possibility of building "a world that is more secure, more prosperous, and more sustainable both economically and environmentally."

In particular, the conference identified four major developments that would have to take place in order to achieve a positive answer to the question that brought them together: (1) *A transition from an exponentially increasing to a stable world population.* (2) *An energy transition to an era where energy is produced and used with high efficiency without aggravating other global problems.* (3) *A resource transition to reliance on nature's "income" and not depletion of its "capital."* (4) *An economic transition to sustainable growth and a broader sharing of its benefits.* Transition, of course, means change, and the conference report outlined an Agenda for Action that gave details on the changes that would be needed if the goals were to be met. Responsibility for action was divided among four entities: government (which had the major role), business, science, and citizens.

This, then, is the environment revolution—in capsule form. Will it be a peaceful transition, or turbulent? Will change be forced on us by events, or will we act before problems reach uncontrollable proportions? These questions, and the issues that prompt them, are part of the new discipline of *environmental science*, which has appeared only recently in the curriculum of most colleges and universities.

## The Perspectives of Environmental Science

Environmental science is simply a new name for an old activity: learning how to live on this planet. Because of the magnitude of our population and the new technologies that we have employed, however, learning how to live on the earth has become an extremely difficult task. The complexity of the issues and multiplicity of problems require that environmental science be an interdisciplinary activity, crossing the boundaries between the natural and social sciences and the humanities.

For the same reason, environmental science is heavily

dependent on *systems analysis* as a strategy for tackling complex problems. Systems analysis is defined as an orderly and logical organization of data, information, and processes into models, and the subsequent testing and work on these models in order to validate and improve them. It is an approach that is ready-made for our computer age. Problem-solving in environmental science, however, is absolutely dependent on knowledge of natural systems, and this is the province of ecology.

### *Ecology: From Local to Global Concerns*

My first impression as a student in a college ecology course was that this branch of biology makes sense out of everything else that I had learned about living things. So I became an ecologist—and those were the days when nobody knew what that meant. With the environmental awareness of the 1960s and 1970s, ecology became a household word. Ecology is the foundational science for understanding how natural systems work; in that sense, ecology provides us with the most basic tool for our stewardship tasks.

The primary working model for ecology is the *ecosystem*, an environmental unit (such as a forest, pond, grassy field) consisting of all of the organisms (plants, animals, microorganisms) and the nonliving parts (soil, water, chemicals), and emphasizing the interactions and interdependencies within the system. Biologically and economically, we are absolutely dependent on ecosystems, and there are now so many of us that it is virtually impossible to find any ecosystem that has not been affected by our activities. Our biological dependence reminds us of our animal needs—food, oxygen, water, waste disposal, shelter. Our economic dependence on ecosystems is just as absolute; in *The Twenty-ninth Day*, Lester Brown describes how four biological systems—fisheries, grasslands, forests, and croplands—form the foundation of the global economic system. These supply all of our food, and all of the raw materials for industry except for minerals and petroleum-based synthetics.

In spite of these obvious dependencies, our economic and technological activites often produce devastating effects on individual ecosystems. But the consequences are a matter of scale. A marsh gets filled, a woodland is converted to building lots, a stream is overwhelmed by sewage plant effluent, and perhaps only a few conservation-minded folks mourn their passing. We are lulled into a false sense of security, content in thinking that there are more such ecosystems just over the horizon. The same sense of security applies to our personal polluting activities: a little pesticide used here, some fuel burned there, a few bags of trash put out every week. Surely these have no broad significance—or do they? Is there not logically some impact of the accumulation of these thousands of small-scale changes? Shouldn't we be thinking about larger-scale effects? These are the kinds of questions addressed by *global ecology*.

Global ecology is an emerging ecological discipline that studies ecological principles and problems on a worldwide basis. The unit of study is the biosphere—that part of the earth containing living organisms. Again the emphasis is on interactions, but the scale has changed by orders of magnitude. Some components of global ecology focus on accumulating local and regional impacts—for example, acid rain, coastal pollution, habitat loss, soil erosion—to determine how and when they reach large-scale importance. Other human impacts can't be seen or dealt with at the local level—chemical changes in the atmosphere, for example—and so require new approaches for study, approaches involving atmospheric physics and chemistry and studies of interactions between the atmosphere, biosphere, and oceans.

By documenting the extent of erosion, deforestation, and the levels of atmospheric gases, for example, the results of these larger-scale studies can provide guidelines for local and regional action. Indeed, ecology has gotten to be serious business. Ecologists have often been in the front lines of environmental battles as they have sought to bring their understanding to bear on the political processes in society. This will be borne out as we look at the four major components of the environmental crisis.

### *Critical Pollution Issues*

Pollution—the introduction of wastes to natural environments—accompanies practically everything that we do. Several forms of pollution occur in such magnitude that they have attracted national and international attention. Acid rain drops out of the skies onto areas at great distances from the source of the acids, and destroys forests and lakes in poorly buffered soils. Many lakes not affected by acid rain are likely to be enriched with waste nutrients (eutophication) to the point where they can no longer be enjoyed because of excess plant growth. Hazardous chemicals are introduced into the environment either intentionally (as with herbicides and pesticides) or through neglect (as with industrial chemical wastes), contaminating food and water and sometimes rendering an environment unfit for life (in Love Canal, New York, or Times Beach, Missouri). Chlorofluorocarbons used in spray cans and refrigerants are accumulating in the stratosphere, damaging and posing a serious threat to the ozone layer, which absorbs ultraviolet radiation (this deterioration causes skin cancers and damages vegetation).

Another very serious and complex pollution problem is the buildup of carbon dioxide in the earth's atmosphere due to the burning of fossil fuels. Much of the carbon dioxide given off when petroleum, gas, and coal are burned remains in the atmosphere. Because of the "greenhouse effect" of carbon dioxide (and other gases we release) on the global heat balance, major global warming appears to be inevitable. Global warming in turn will affect sea level as ice caps are melted in polar regions, and the expected sea level rise will produce major coastal flooding with unprecedented effects on population distribution. This scenario is based on extremely reliable scientific reasoning and model-building. The gradual nature of the carbon dioxide buildup and its direct link to energy generation make this one of the most difficult problems imaginable.

The Council on Environmental Quality, a watchdog governmental agency, produced a report in 1981 entitled *Global Energy Futures and the Carbon Dioxide Problem.* In the preface of that report, Council chairman Gus Speth (now chairman of the World Resources Institute) issued an eloquent call for stewardship:

> Whatever the consequences of the carbon dioxide experiment for humanity over the long term, our duty to exercise a conserving and protecting restraint extends as well to the community of plant and animal life existing with us. There are limits beyond which we should not go . . . Although our dominion over the earth may be nearly absolute, our right to exercise it is not. People have altered the face of the planet throughout history, but the power of today's technology and our growing capacity to foresee the possible consequences of our acts puts us in a new moral position. The responsibility for the carbon dioxide problem is ours—we should accept it and act in a way that recognizes our role as trustees of the earth for future generations.

Speth makes pollution a matter of morality. Without being explicit, he places the issue of the global carbon cycle directly in the context of our God-given responsibilities for the creation. In effect, *pollution is a matter of sin.* If we know that what we are doing is harmful, and we can prevent it, then we are committing sin if we do not take action. And sin, as we well know, has its consequences. In this case, the consequences are enormous. It is clear that coping effectively with pollution at all levels—local to global—is a major facet of the environmental revolution.

### *The Human Population: Growth, Health, and Control*

How many people can our planet support? Putting it in ecological terms, what is the earth's carrying capacity for our species? This is not a simple question. There are vast differences between different nations and the biological resources on which they depend. We also could ask, how many people *should* this planet support? Would we want to live on an earth where the major goal is to support as many people as possible? As we will see, this would take an enormous toll of other species, and would undoubtedly impoverish the quality of life for most of humanity. *Demographers* inform us that the world population recently passed 5 billion, and continues to increase at a rate of 1.8 percent per year. These figures cover up a basic fact, however: We live in a "demographically divided world," as a recent Worldwatch Paper by Lester Brown and Jodi Jacobson stated:

> Our contemporary world is being divided in two by demographic forces. Nearly half the world, including the industrial countries and China, has completed or nearly completed the demographic transition. These countries, where fertility is at or below the replacement level, have either established a balance between births and deaths or they are in the process of doing so. In the other half, where birth rates remain high, rapid population growth is beginning to overwhelm local life-support systems in many countries, leading to ecological deterioration and declining living standards."

The size, geographic distribution, and growth rate of the human population connects with every other facet of the environmental crisis at the most fundamental level. The half of the world in which population growth is now slow or stopped (North America, Europe and Soviet Russia, eastern Asia) can come to grips with their pollution and resource problems, and can anticipate improving living conditions. But the other half of the world, where population is increasing at an average rate of 2.5 percent per year, represents regions that are now in their fourth decade of rapid population growth (Southeast Asia, Latin America, Indian subcontinent, Middle East, and Africa). In many of these regions, living conditions are deteriorating because of food and fuel-wood scarcities, overgrazing of livestock, soil erosion and slow economic growth. Yet demographers project population increases of an additional 5 billion in these regions before they stabilize. Brown and Jacobson call these "projections of disaster" because of the inability of the life-support systems of these regions to sustain such populations.

Population problems in the United States are of a different nature. We have yet to deal effectively with immigration, both legal and illegal. In several decades, immigration will represent the major source of population increase, if current trends continue. We can no longer be the safety valve for political and economic problems occurring throughout the world, and welcome all refugees who seek a new start on life. Yet we are a nation of immigrants and their descendents, and we have a strong commitment to the principle of immigration. Another set of problems we do not handle well is signaled by the high number of births to unmarried teenagers, the rising number of children born out of wedlock, and the high number of abortions. Our society—which uses sex to sell everything—has failed to communicate the meaning of sexuality and sexual responsibility to younger generations (problems we considered in chapter 10). Finally, a case can be made that in terms of resource use and pollution, every additional American is the equivalent of ten people from the third world!

World health problems are divided along the same demographic lines as population growth. The most serious health problems in the low growth rate and more industrialized countries are the chronic, noninfectuous diseases (cancer and cardiovascular problems), and the lifestyle problems of smoking, alcohol, and drug abuse. In the high fertility nations, infectious diseases like tuberculosis, malaria, pneumonia, dysentary, and worm infestations still account for most of the mortality and sickness. The growing epidemic of AIDS threatens to become a major factor in mortality in many of the less-developed regions. And in some of these nations, chronic malnutrition and hunger continue to be serious problems—in spite of the relief efforts of many donor countries.

The evidence points to the obvious need on the part of many of the world's most populous nations to bring population growth to a halt before their carrying capacity is exceeded. This is a task of heroic proportions; in recent years, only mainland China has succeeded in bringing about a rapid decline in fertility, and had to resort to methods that would be unacceptable in most free societies. The ethical problems raised are enormously difficult, yet the option of taking no action at all would appear to raise even greater problems.

Several key steps form the basis of a population policy that will reduce growth: (1) *Make family planning services available to all women in the society*; (2) *because couples in many countries still want large families, educate people about the consequences of high fertility*; (3) *since steps 1 and 2 will not bring about a sufficiently rapid decline in fertility in most countries, additional steps must be taken by governments to bring pressures to bear that will lead to less childbearing*. These involve economic incentives to keep families small, such as rewards for being sterilized and tax deductions limited only to the first two children (many more strategies are available—some of them highly controversial and probably useless, like giving out two tickets to each couple, which entitles them to two and only two children!). Failure, according to Brown and Jacobson, "could be catastrophic. The issue is how—not whether—population growth will eventually be slowed. Will it be humanely, through foresight and leadership, or will living standards deteriorate until death rates begin to rise?" The transition to a stable world population is clearly one of the most fundamental components of the environmental revolution.

### *Resource Limits: Focus on Food*

Of all the resources we take from the earth, food is the most vital. Undoubtedly, access to food is one of the most basic human *needs*—but is it a human *right?* The sale of foodstuffs represents a major source of income for many nations; should food be given away? What factors are responsible for the hunger and malnutrition that stalk many of the poorer countries? These questions suggest that world hunger and food supply is one of the most complex issues of environmental science.

Pictures—and now videos—of malnourished African children can disturb us enough to send a pledge to relief agencies, but most observers feel that food aid has only encouraged a deeper food dependency on the part of the poorer nations. Food aid—as necessary as that is—only alleviates the symptoms. In the majority of cases, *the root cause of hunger and malnutrition is poverty.* Food flows in the direction of economic demand, not need. Most malnourished and hungry people lack both the means to grow their own food and the means to purchase it. These people frequently live in a land where desperate poverty and wealth exist side by side. They most certainly live in a world where wealthy and economically powerful nations exist side by side with desperately poor nations.

It is clear that overcoming world hunger and malnutrition requires addressing the root causes rather than simply donating food. Although much of the aid from rich nations is targeted toward agricultural and economic development, there is far too little of it and it rarely addresses the major obstacle in the way of development: economic domination of poorer nations by the richer ones. But it is equally true that alleviating the most devastating poverty means changing the political and social system that controls land and wealth and distributes services in the society—an unpopular agenda.

In some of the poorest nations, individual poverty is symptomatic of conditions in most of the nation. Many sub-Saharan African nations simply lack the land resources to feed their rapidly increasing population, and for at least a decade these nations have experienced a steady decline in per capita food production. Many observers believe that major areas of the African continent have been so impacted by human and livestock use that the entire continent is in a state of ecological decline—the carrying capacity has already been exceeded. The only deterrent to hunger and starvation is food aid, and all too often this is mismanaged and insufficient.

*Meeting the challenges of world hunger* requires several critical—and simultaneous—developments:

1. *A far greater commitment on the part of both donor and recipient nations to meeting the needs of the rural poor;* this will undoubtedly mean diverting aid and national resources from military and showcase industrial projects. The ultimate goal should be self-sufficiency.
2. *Reducing population growth from the range of 2 percent to 3 percent per year to below 1 percent per year.* China's experience suggests that this can only be accomplished by centralized and intensive programs to lower fertility.
3. *A greater commitment by the wealthy nations to establishing a system of world food security, where the fluctuations in food production caused by local and climatic factors can be compensated by food stored for just such purposes.*
4. *Even though public and private organizations are contributing some $35 billion per year for development and relief aid, more is required.* One obvious source is the United States, whose foreign aid pattern has increasingly shifted over the last decade from development assistance

to military help, and whose contribution to aid is now the smallest among the developed nations in terms of percentage of gross national product.

Many other resources, both renewable and nonrenewable, are taken from the earth to support human activities. The record of our management of these resources is not encouraging; we tend to use the nonrenewable resources as if there were no future, and we have so mismanaged the renewable resources that in many cases the productive base is almost eliminated. The environmental revolution will require that we manage resources more wisely, or else they simply will not be available for us or future generations.

### *Biological Depletion*

The foregoing problems—pollution, population growth, and resource use—all have devastating effects on the plant and animal species that share this planet with us. It is all too easy to document the impacts of air and water pollution on human health and ignore their effects on the rest of the living world—unless, as with trees and acid rain, we wish to harvest them. Nothing is more destructive to natural habitats than the activities associated with growing plants and animals for our uses. As the human population increases and occupies more and more of the remaining natural land, there is a diminishing of the habitats that are required to support the natural species, and eventually they are eliminated—just as surely as if we were bent on exterminating them. Our very use of the term "natural resources" suggests our real attitude about the other species—they are viewed primarily in terms of their economic value to us and not accorded any intrinsic value.

Nowhere is this approach more evident than in the tropical rain forests. At the current rate of population growth and forest clearing, the Latin American rain forests will have shrunk to half of their original size by the end of this century.

These forests are home to an amazing diversity of plants and animals. They also comprise the bulk of the world's living carbon biomass, which, if released into the atmosphere, would seriously raise carbon dioxide levels. The tropical forests are being cleared and burned both for their timber and for cattle ranching and cropping. Writing for Worldwatch Institute, Edward Wolf estimates that 12 percent to 15 percent of the plants and birds of the Latin American forests will be eradicated by the year 2000; and if forest clearing continues so that the only intact stands are those in protected areas, two-thirds of the fauna and flora could be lost. This is a loss of 60,000 species of plants alone. The process of massive forest clearing is under way; there is nothing to suggest that it will not duplicate the experiences in parts of Southeast Asia and West Africa, where 10 percent or less of the original forests remain.

The loss of species is not confined to the tropical forests. Wolf explains that "In North America . . . national parks considered the last refuge for some of the continent's most distinctive wildlife are proving inadequate to the task." He speaks of a faunal collapse that is underway as we increasingly constrict the sizes of ecosystems that sustain vulnerable species. The worst-case scenario for biological depletion—reduction of forests to currently protected areas and conversion of the remainder to agricultural use—gives us a picture of a coming world with greatly diminished biological diversity, where the dominant plants and animals are the weeds, cockroaches, rats, and pigeons. Gone forever will be millions of species of plants and animals, each of which represents a unique page in the book of life. It is literally true that a "book-burning" is under way.

How can this process be stopped? Given the continued population growth in the Third World and the economic pressures to convert natural areas into money, this is a hard question. Perhaps the key to an answer lies in recognizing the rights of other species to exist as well as the potential for those species to be of value to us. It continues to be true that natural ecosystems are life-support systems for us; how much can we lose before our own existence is threatened? All of our food and natural fiber and many useful chemicals and drugs come from species taken from nature; we lose unknown future potential when we drive species to extinction.

We need ecological knowledge of the crucial size limits of tropical and temperate ecosystems for sustaining species diversity. We need a determination by political bodies to maintain natural areas against the economic pressures that threaten them. And we need the wisdom of a new enterprise in the academic world—*restoration ecology*. This is a discipline that attempts to put ravaged ecosystems back together, collecting the key species and carefully managing their establishment on land they once occupied.

The environment revolution, when it occurs, can take one of two directions: either we live in an impoverished world where our activities have led to mass extinctions—and suffer the consequences—or we live in a world where biological diversity still exists, where we can point with satisfaction to our stewardship of the other species God has placed on earth.

### Addressing the Issues

It should by now be clear that our analysis of the environmental crises places us in sympathy with the message of the Jeremiads. Things indeed are not going well; one would have to be uninformed or insensitive to be complacent about this array of problems. The environmental revolution is upon us; the responses of our civilization to it will reveal as never before what having dominion means. Two questions emerge as critical: *What needs to be done? Why should we do it?* If these can be answered, the task of how to bring effective action is greatly facilitated. To illustrate the significance of the first question, we will consider the problem of acid rain. As we do, it will become clear why it is also important to answer the second question.

Ecologists have shown clearly that acid rain affects lakes by overcoming the natural buffering capacity of the water; the devastating effects of acidification on fish and other natural species in the lakes have been well documented. Acid rain also contributes to the destruction of forest trees. Environmental scientists broaden our knowledge of acid rain by tracing it to its origins in the sulfur oxides released when coal is burned at a complex of power plants in the midwestern United States. (There are other sources, but this is the major point source).

The following statements seem obvious: (1) It is wrong to destroy lakes and forests; (2) it is therefore wrong to release large amounts of sulfur oxides into the air, since the consequences are dead lakes and forests. These are ethical judgments based on ecological information. But the law allows the midwestern power plants to release the sulfur oxides, so in another sense it is not wrong for them to do so. Their right to pollute supercedes any morality derived from our ecological knowledge. This is a legal judgment. We should add here that all new conventional power plants are required by law to install stack scrubbers, which remove most of the sulfur oxides. In this case, the law discriminates between power plants; the older plants are allowed to pollute because they are economically important; retrofitting them with stack scrubbers is judged to be too costly.

Let's probe a little further. We all drive automobiles that release nitrogen oxides into the air. Environmental science has shown that one-third of the acids in acid rain can be traced to the nitrogen oxides released primarily by the burning of fossil fuels in internal combustion engines—our cars, trucks, and buses. So all of us contribute to acid rain, but our individual contributions are quite small, and they are allowed by the law. If we built a new power plant, however, we would have to install stack scrubbers. This points to an interesting dilemma: Actions that are considered moral (or appropriate, at least) on a small scale may not be considered appropriate on a larger scale. Other examples could be cited: the cutting of trees, the destruction of natural habitat (which impacts wildlife as surely as killing them directly), and so forth. The difficulty is in drawing lines; at what point should some pollution or environmental change become wrong? What shall be allowed, and what shall be prohibited? How do we make such decisions?

Clearly, we must first determine what needs to be done. Here is where ecological knowledge and environmental science must be brought to bear on the problem. In most cases, the cause-and-effect relationships involved in environmental issues are well enough understood so that corrective action could be taken. Why, then, is it not taken? *Because we are not willing to pay the cost.* There are costs associated with every unresolved environmental issue. The costs lie primarily in the arenas of economic value and human freedom of action. So our second question can be restated: Why should we give up some economic advantage or some desired freedom of action in order to resolve an environmental problem? The compelling answer to this question is, *we do so because it is right, because it is just.*

### *That Justice be Done*

Justice is the ethical principle that insists on fairness, on the guarantee of rights, on seeing to what is deserved. We are quite comfortable with the notion of human rights, even though there may not be complete accord on what those rights are. In any well-ordered society, human rights—once defined—are protected by the law.

Matters of justice and human rights are woven into many of the issues we have raised. For example, do we accept a human right to food and other basic sustenance needs? Some societies do, some don't. Is there any sense in which justice is applied to the distribution of resources between nations? We know that God is concerned with justice, and he expects his people to act with justice (see Amos 5:21–24, or Micah 6:6–8). Ron Sider, in *Rich Christians in an Age of Hunger*, attacks these problems from a scriptural perspective, and challenges Christians to respond to the injustices around them and work towards a resolution of hunger-related issues.

But what about justice for the land, or justice for other creatures? Does the land have rights? Do the animals and plants have rights? In other words, is there an *environmental ethic?* If there is such an ethic, is it being translated into law? Can we see it in our attitudes? If we can establish the rights of the land and the creatures, then we have our basis for answering why we should do what needs to be done.

### *Environmental Ethics*

Ethicist and naturalist Holmes Rolston III is a pioneer in the developing discipline of environmental ethics. His writings in *Philosophy Gone Wild* are a rich and eloquent statement to the effect that we do indeed have an environmental ethic, but it is not yet where it needs to be. Rolston points out that much of our ethical thinking about the environment is *anthropocentric*—our motives are tied in with human values, needs, and desires. The things in nature do not have value apart from their importance to humankind. For example, this ethic would say that we must strive to maintain stability in ecosystems because they are our life-support systems. Implicit in that reasoning is the moral position that human life ought to be preserved.

Most environmental ethical reasoning fits into this category. This kind of ethic, however, leaves openings for serious conflict with other human values, with no clear pathway for resolution. What happens when job-producing economic activities threaten endangered species or whole ecosystems? The fate of the tropical forests tells us the answer. However, the anthropocentric environmental ethic is far better than none at all.

Rolston argues convincingly for an environmental ethic that is independent of human rights and values, one that recognizes rights as intrinsic to the natural world. This is the *naturalistic ethic*. Endangered species are to be morally considered in their own right. Some landscapes—entirely apart from human interests—have a right to exist, and we have a duty to preserve and protect those landscapes. The ethical basis for such reasoning and the emerging ethical principles are not yet well established. Yet, there are signs of movement towards such an ethic, as in the Endangered Species Act passed by Congress. Rolston speaks to our duties to endangered species:

> Several billion years worth of creative toil, several million species of teeming life, have been handed over to the care of this late-coming species in which mind has flowered and morals have emerged. Ought not those of this sole moral species do something less self-interested than to count all the produce of an evolutionary ecosystem as rivets in their spaceship, resources in their larder, laboratory materials, recreation for their ride? Such an attitude hardly seems biologically informed, much less ethically adequate. Its logic is too provincial for moral humanity. Or, in a biologist's term, it is ridiculously territorial. If true to their specific epithet, ought not *Homo sapiens* value this host of species as something with a claim to care in its own right?

It would be inappropriate to leave this subject without suggesting the possibility of a Christian environmental ethic. The 1987 Au Sable Institute Forum addressed the subject of "A Christian Land Ethic"; some critical groundwork for a *theocentric environmental ethic* was laid by the participants. In particular, ecologist Susan Bratton demonstrated the scriptural support for preserving wild nature. For example, as God declared his creation of both wild and domestic-type animals good, this established wild nature as having *value* and worth in its own right. The command to have dominion—described as a stewardship—implies that humans have a responsibility to care for the other creatures. Later passages in the Bible (for example, Psalm 104) indicate God's provision for wild things and praises God for arranging things so wisely. Bratton concludes:

> With a careful reading of the Scriptures, we find numerous grounds for the protection and preservation of wild nature. Nature has a number of values which exist independent of human interest and need. . . . The best means of determining how much land must be set aside and how much habitat should be left for wild species is to investigate diversity, productivity, ecosystem integrity and beauty, and to determine when critical losses will occur and when systems can not be replaced. As part of a Christian land ethic, sites should be left for wild species and the harvest or displacement of wild species should be managed in such a way that losses do not occur.

## Developing a Position: A Paradigm

Finding the way to a resolution—at least a resolution in principle—is clearly not going to be an easy task. The issues are complex, interrelated, and numerous. I would like to conclude this chapter by offering a paradigm for doing some of this hard work. The intent is to help track a path to a position that could be considered both responsible and Christian. To take things from a position to a policy is a quantum leap, but the position is essential if we are ever to get off the ground!

1. *The knowledge base*—it is important to have accurate and adequate information; the complexity and interdisciplinary nature of the issues makes this a formidable task, in some issues. In most cases, however, there is a sufficient knowledge base to define the ethical parameters. This is the task of environmental science, founded on ecological information.
2. *Predominant attitudes*—these refer to the predominant thought patterns of a society, which, in the absence of

strong legal and/or ethical constraints, determine how people will order their relationships. These very directly reflect the dominant world view of that society. They are also very difficult to influence. The predominant attitudes can usually be determined by examining (a) the political and legal components of an issue; and (b) the common practice in society related to the issue as revealed by the voluntary acts of people. Because of their dominant role in human activities, economic considerations represent a fruitful place to start.

3. *Ethics*—right and wrong must be defined (ethical rules, in other words), which suggests that ethical principles and an ethical base must also be articulated (see chapter 10). We must expect different normative ethics for human relationships, economic relationships and environmental relationships; these will sometimes (might I say, often) conflict, in spite of the fact that humans (and economic activities) are dependent on ecological systems. We should expect different ethics for relationships within a given society and for relationships between societies or nation-states. And we must consider whether there is a distinctly Christian ethic that applies to the issue.
4. *Legal arena*—what are the existing laws and regulations that apply to the issue? What ethical principles and rules do they reflect? What ethical concerns are not dealt with? Are the laws and regulations effective, or should they be strengthened or changed? How did they come into existence?
5. *Resolution*—a judgment must be rendered. Assuming that the issue in question is not resolved, and in fact might be worsening, what will it take to turn the problem around and move in the direction of resolution? What attitudes must be changed, and how can this be done? What ethical principles must be articulated and built into

a consensus consistent with the limitations of our pluralism? To what extent are legal and regulatory changes needed? How will this be accomplished? Of great importance, what are the implications for other issues of resolving or not resolving this issue.

In my course "Stewardship and Survival" at Gordon College, students are given an assignment to choose an issue related to the course—it can be local, regional, or global—and put this paradigm to use in developing a Christian position on that issue. The outcome is a six- to ten-page paper—and, from the students' perspective, a struggle. However, they agree that there may be no better way to deepen their understanding of what it means to do stewardship work.

## Transition to a Sustainable World

The environmental revolution will be part of your future, and the future of your children. I do not doubt that enormous changes are coming—we must learn to live within the carrying capacity of the biosphere, and we must do it with justice. I would like to share with you the vision of the future from *The Global Possible;* it is a vision I can subscribe to, and in the next chapter, I will explore with you the Christian dimensions of that vision:

> As we consider the contours of a brighter and sustainable future, its features become clearer. World population is stabilized before it doubles again, and the erosion of the planet's renewable resource base—the forests, fisheries, agricultural lands, wildlife, and biological diversity—is halted. Societies pursue management practices that stress reliance on the "income" from these renewable resources, not a depletion of the planet's "capital." Enlarging this income requires sophisticated management and more intensive use of prime farm and forest lands and fisheries, as well as the application of new technologies to improve agricultural yields, control pests and spoilage, and exploit new opportunities such as aquaculture, hydroponics, and

salt-tolerant crops. People and resources are protected from the costly consequences of pollution and toxification and from disruptive climate change. Human activity becomes more "closed" in the ecological sense, so that it does not impair the functioning of natural systems. Manufacturing processes produce less waste, and what waste is produced is reused in other processes. Advanced technologies are widely applied to achieve high efficiencies in the use of energy and in its production from solar, biomass, and other renewable sources. Broadly-based economic growth proceeds in ways that lessen the gap between rich and poor both within and among countries, and the door is increasingly opened to artistic and cultural pursuits in a world where the hard labor of survival is lessened.

## SUMMARY

The *environmental crisis* is brought on by our swelling population and its apparent inability to come to grips with the limits imposed by the environment. The *environmental revolution* is a future revolution that will come as humankind is increasingly forced to live within the sustainable limits of the earth. The major components of the environmental crisis are *pollution, population growth, resources, and biological depletion.* Coping effectively with these major problem sources requires meeting a number of goals, all directed toward achieving a sustainable relationship between the earth's people and the land.

Environmental science is an interdisciplinary activity that studies the interactions of humans with the environment. It is based on ecology, which seeks to understand how natural systems work from local to global levels. The approaches of these related disciplines are applied to the four major components of the environmental crisis. Each is sketched out in general terms, and suggestions are given for the direction that must be taken if the environmental revolution is to be a rational process of accomodation to a sustainable earth.

Two *critical questions* emerge as we face the coming

changes: *What needs to be done? Why should we do it?* The example of acid rain shows the linkage between these questions. In most cases, we do not take corrective action because we are not willing to pay the costs. We usually know what needs to be done, but we do not want to forego some economic advantage or some freedom of action in order to take corrective action. It is suggested that taking such action is a matter of *justice*.

Justice in connection with human rights is a familiar principle in our ethical interactions, even though it is not uniformly followed. We are less familiar with justice for the land and other creatures, the subject matter of environmental ethics. Several *bases for environmental ethics* exist: *anthropocentric, naturalistic, and theocentric*. Although most of our existing environmental ethic is anthropocentric, this is better than no ethic. There is a strong need to develop a deeper ethic based on values external to humans, however, if we are to resolve many of the most difficult environmental problems.

A paradigm is offered to track a path to a position on a given issue that is both responsible and Christian. It begins with the *knowledge base*; explores the *predominant attitudes* in a society; uncovers the *ethical questions* raised by the issue; looks at the *legal components*; and then comes to a *resolution*. This paradigm can help to reveal what it means to do stewardship work.

The environmental revolution will involve enormous changes in taking us from where we are now to a sustainable future. A glimpse of this future is provided by a quote from *The Global Possible*.

*Chapter 13*

# WHEN EARTH AND HEAVEN ARE ONE

This is my Father's world,
O let me ne'er forget
That though the wrong seems oft so strong,
God is the ruler yet.
This is my Father's world:
The battle is not done;
Jesus who died shall be satisfied,
And earth and heav'n be one.

"MY FATHER'S WORLD," BY MALTBIE BABCOCK

## Worldview Crisis

In their book *The Transforming Vision*, Brian J. Walsh and J. Richard Middleton suggest that our society is in the throes of a worldview crisis. Before we explore this important concept, let's review what we have learned about worldviews.

In the first chapter, we defined worldview as a comprehensive framework of beliefs that helps us to interpret what we see and experience and also gives us direction in the choices that we make as we live out our days. Every person has a worldview, acquired primarily from the surrounding culture; the culture itself displays a communally shared, dominant worldview that determines the direction taken by our social, political, and economic arrangements.

*Naturalism* is a worldview that claims there is nothing beyond the material world, that all of nature is governed by autonomous "natural laws" that operate strictly on the basis of cause and effect. There is no direction or plan to events past, present, and future; the concept of God is irrelevant to the natural world and, by extension, to human life. Naturalism can be expressed in philosophical terms, and it can also be unconsciously incorporated into people's thinking.

In opposition to naturalism is *theism*, which holds that God is the central fact of existence; that even though the natural world obeys "natural laws" and exhibits cause and effect behavior, God upholds and directs the world by his creative and sustaining activity. Nature is not autonomous; its lawful behavior is the result of obedience to the God who created the whole natural order.

Our look at the scientific enterprise introduced us to the importance of *shaping principles*, those values, assumptions, beliefs, and commitments that scientists bring to their work. Worldviews are highly important shaping principles that play an important role in controversies involving science and faith. They are rarely revealed in the published work of scientists, and they often reflect poorly examined philosophies. As we began to deal with the encounter between biology and Christianity, we quickly found competing worldviews.

Historically, the worldview of Darwin's time embraced natural theology, which saw design and purpose in the natural world and ascribed them to God's wisdom. Applied to biology, natural theology included an explanation for the origins of adaptations and assumed that these were bestowed by God to organisms during the creation week. As such, natural theology was a form of *deism*, where God got things started and then conveniently removed himself from the scene. The Darwinian revolution otherthrew this worldview by providing a new paradigm, where organisms and adapatations originated from within the natural world. At the same time, *agnosticism* appeared as a potent naturalistic worldview.

We took some pains to sort out the scientific claims of evolution from its worldview extensions (*evolutionism*), and found that this distinction was vital to gaining insights into the creation/evolution controversy. We also saw the strong influence of evolutionistic thinking on issues involving the origin of life and the appearance of human beings. And we concluded that a theistic worldview does not require rejecting the biological dimensions of evolution—indeed, the deep issue is Creator versus no Creator.

The chapters on stewardship and the biomedical, genetic, and environmental revolutions shed light on some of the dominant worldview components of our modern society. In particular, we encountered *secularism*—the separation of religion from human culture as a result of the belief that people and nature are autonomous; and *materialism*—the belief that our highest good consists of satisfying our desires and needs through material prosperity. These dynamics have strongly influenced the development of modern social, political, and economic life; they have also determined the ways in which we use natural resources and relate to natural environments. We reviewed some ethical questions that have emerged as a result of new biological knowledge—and documented the appearance of bioethics as an application of ethical and moral reasoning to such varied fields as molecular genetics, medicine, and the environment. Finally, we encountered the concerns of justice for people and for the land and other creatures, as we asked what to do about the different issues of the environmental crisis. We saw that worldviews play an important role in the attitudes and ethics being brought to bear on these problems.

In summary, we have seen strong interaction between worldviews and scientific thinking; worldviews have played a vital role in the issues that bring religion and science into contact. Naturalistic worldviews continue to guide the views of most people in science, and are unconsciously incorporated into the dominant worldview of our modern Western culture. We have also seen that worldviews play a crucial role in the political, social, and economic arrangements of our culture.

The ethics and morality of our culture reflect the workings of secularism and materialism far more than they do religious beliefs; in a real sense, these have become substitutes for the religion much of our culture has "outgrown."

Now, however, our worldviews seem to be crumbling. Our "idols"—those things or forces in which we have placed our faith—have failed us; our strong belief in the capabilities of science and technology, our embracing of economic prosperity have led us to the serious social and environmental problems that we documented in the last chapter. Our efforts to dominate nature and society so as to satisfy our desires, amplified now that there are so many of us, have only served to intensify the sense that things are getting out of control. We look with apprehension to the future, wanting very much to believe the Cornucopians and their message that all is well, yet suspecting that we are living on borrowed time and will soon have to cope with an unprecedented set of changes—the environmental revolution.

We cannot be content with the way things are going, yet we seem unable to fix our sights on any clear path to the future. The dominant worldview of Western societies has not served us well. When we look to other cultures, we frequently find that their traditional worldviews have been swamped by the internationalization of Western culture. They have largely adopted Western values, and now are facing the same set of problems that we are experiencing, except that they have even less of a resource margin for the future.

Our early excursion into worldviews also affirmed the importance of developing a Christian worldview; it is the duty of all Christians to develop and apply a worldview that faithfully reflects God's truth. An important component of a Christian worldview is the belief that Scripture speaks vitally to everything in our life and world, that there are no compartments labeled "sacred" and "secular." Such a worldview will guide us in interpreting the issues raised in this book: origins, medical and genetic concerns, population and resource use, pollution, justice, and so on. Most important, a biblical worldview will give direction to our responses to these issues.

The biblical worldview also has the key to the worldview crisis. It brings a prophetic word to our culture, a vision for life as it should be, and an agenda for realizing that vision. In short, the biblical worldview informs both Christians and those outside the faith about God's concerns and his program for the future. It enlists us in the accomplishment of three vital tasks: *proclaiming the good news of salvation*, *reforming the culture*, and *redeeming the Creation*. To put form and function to this vision, we must first return to the beginning of things and briefly trace them through history.

## The Flow of Time: God's Plan

Let us review the biblical picture of beginnings, as documented in early Genesis. By Word and Wisdom, God created the heavens and the earth and populated them with an amazing array of living creatures. Although all of creation glorifies God and was declared by him to be good, God made one species—humankind—in his image, and gave them dominion—rulership—over the rest of creation. This dominion included the responsibility to develop a culture ("subdue the earth") and to care for the creation as his representatives—stewardship. But the first people—Adam and Eve—were led by Satan into sin; they rebelled against God. Doubting God's word to them, they asserted their freedom and reached for forbidden knowledge.

This act, highly symbolic in its Genesis description, brought profound consequences: Sin would continue to plague us and deeply mar our ability to image God, and through our sin the rest of creation would be affected. In particular, our sin alienated us from God, from our fellow human beings, and from the rest of the created order. We speak of the world as a "fallen world," and rightfully assign to the Fall all of the misery, pain, and suffering that continue to afflict our race. Most of this is traced directly to the effects of sin on the way in which we have carried out the cultural mandate. Not only are individuals sinful, but the very structures of society also have incorporated the results of human sin.

There is good news, however: God has not abandoned his works of creation. The Bible speaks to us of God's redemptive concern for humankind, and we saw that this concern extends to all of creation. The redemptive work of his Son, Jesus Christ, makes possible the reconciliation of the relationships broken by the Fall. That which was broken can be made whole. God invites us as individuals to come over to his side—to be reconciled. And as his stewards, we are privileged to participate in redeeming the creation. This task has two major dimensions: a reforming work, and a healing work. The culture must be reformed—brought into conformity to God's normative law; and the creation must be healed—restored to ecological wholeness. When Jesus came, he announced the arrival of the Kingdom of God. God's Kingdom is under way, and the reforming and healing work we are called to is identified as Kingdom work.

Finally, as we trace the flow of time into the future, the Bible speaks of a time when peace will reign and justice will prevail, a future when the creation will be purified of its corruption. At this time, the Kingdom of God will be in its final and intended manifestation, and the creation will be whole.

### *The Kingdom of God*

Throughout the Old and New Testaments, the theme of the Kingdom of God runs as a unifying strand. The Kingdom refers to the rulership of God over the cosmos, which in the end will see all things and all people living in peace and wholeness, experiencing all of the goodness of God and none of the effects of sin. Many are the references to the Kingdom in Scripture; one good example is in Psalm 145:10–13:

> All you have made will praise you, O Lord; your saints will extol you. They will tell of the glory of your kingdom and speak of your might, so that all men may know of your mighty acts and the glorious splendor of your kingdom. Your kingdom is an everlasting kingdom, and your dominion endures through all generations.

In his book *A Kingdom Manifesto*, Howard Snyder summarizes the views of several writers in pointing out the centrality of Kingdom teaching:

> The kingdom is such a key theme of Scripture that Richard Lovelace can say, "The Messianic Kingdom is not only the main theme of Jesus' preaching; it is the central category unifying biblical revelation." And John Bright comments, "The concept of the Kingdom of God involves, in a real sense, the total message of the Bible. . . . To grasp what is meant by the Kingdom of God is to come very close to the heart of the Bible's gospel of salvation." As E. Stanley Jones wrote over four decades ago, Jesus' message "was the Kingdom of God. It was the center and circumference of all He taught and did. . . . The Kingdom of God is the master-conception, the master-plan, the master-purpose, the master-will that gathers everything up into itself and gives it redemption, coherence, purpose, goal."

In his miracles and teaching, Jesus not only proclaimed the arrival of the Kingdom, he also demonstrated what the Kingdom was to be like. When he sent out the twelve disciples into the towns and villages, he instructed them: "As you go, preach this message: 'The kingdom of heaven is near.' Heal the sick, raise the dead, cleanse those who have leprosy, drive out demons. . . ." (Matt. 10:7, 8). These miracles (all of which were performed by Jesus), were miracles of restoration—to health, to life, to freedom from demonic slavery. In the Kingdom of God, there is wholeness.

A key passage is found later in Matthew, where Jesus was admonishing the Pharisees because they accused him of driving out demons with power derived from Beelzebub, the prince of demons:

> If Satan drives out Satan, he is divided against himself. How then can his kingdom stand? . . . But if I drive out demons by the Spirit of God, then the kingdom of God has come upon you. . . . how can anyone enter a strong man's house and carry off his possessions unless he first ties up the strong man? Then he can rob his house. (Matt. 12:26–29).

The Kingdom of God comes in opposition to another kingdom; a struggle is under way. Jesus was here referring to his central role in the struggle. In his life, death, and resurrection, he delivered a mortal blow to Satan's attempts to rule over God's creation; he inaugurated the coming of God's Kingdom into history.

As signaled by the prayer Jesus taught his disciples to pray ("Thy Kingdom come . . ."), the Kingdom of God has not yet arrived in its complete form. There is more to come. The day will come when sin and Satan's hold will be completely abolished. John's vision of this time is recorded in Revelation 11:15 and 21:1–4:

> The seventh angel sounded his trumpet, and there were loud voices in heaven, which said: "The kingdom of the world has become the kingdom of our Lord and of his Christ, and he will reign for ever and ever." (Rev. 11:15)
>
> Then I saw a new heaven and a new earth, for the first heaven and the first earth had passed away, and there was no longer any sea. I saw the Holy City, the new Jerusalem, coming down out of heaven from God, prepared as a bride beautifully dressed for her husband. And I heard a loud voice from the throne saying, "Now the dwelling of God is with men, and he will live with them. They will be his people, and God himself will be with them and be their God. He will wipe every tear from their eyes. There will be no more death or mourning or crying or pain, for the old order of things has passed away." (Rev. 21:1–4)

This is the state of things referred to in Old Testament prophecy as God's *shalom*, a concept to which we now turn.

### *Shalom*

The Kingdom of God is presented as a kingdom of peace. In Howard Snyder's words:

> The Old Testament teaches that God's plan is to bring a universal peace (shalom) to his creation. This means more than the absence of conflict and immensely more than "inner peace" or "peace of mind." In the Old Testament sense, shalom might be called an ecological concept. It carries the sense of harmony, right relationship and the proper functioning of all elements in the environment. . . . In the Old Testament peace is decidedly a this-worldly concept, grounded in the very physical nature of God's creation. It is harmony and wholeness. In bringing peace, God brings healing . . . [see Isa. 11:6–8 and Jer. 33:6]

In his vital work *Until Justice and Peace Embrace*, Nicholas Wolterstorff points to the important connection between Old Testament shalom and the coming of the Messiah:

> That shoot of which Isaiah spoke (see Isaiah 11:1–2) is he of whom the angels sang in celebration of his birth: "Glory to God in highest heaven, and on earth his peace for men on whom his favor rests" (Luke 2:24). He is the one of whom the priest Zechariah said that he "will guide our feet into the way of peace" (Luke 1:79) . . . He is the one of whom Peter said that it was by him that God preached "good news of peace" to Israel (Acts 10:36). He is the one of whom Paul, speaking as a Jew to the Gentiles, said that "he came and preached peace to you who were far off and peace to those who were near" (Eph. 2:17). He is in fact Jesus Christ, whom Isaiah called the "prince of peace" (Isaiah 9:6).

This peace Jesus has come to bring is the peace of reconciliation—of the healing of broken relationships, as shown in Colossians 1:19–20: "For God was pleased to have all his fullness dwell in him, and through him to reconcile to himself all things, whether things on earth or things in heaven, by making peace through his blood, shed on the cross." And as this passage notes, this healing extends to all of creation. Harmony, wholeness, health—shalom—will be restored as a result of the work of Christ.

This, then, is the Kingdom significance of peace. God's Kingdom is marked by peace; Jesus inaugurated the Kingdom, demonstrated peace as wholeness in his ministry, and then in his death and resurrection provided the means whereby peace could be restored to a creation that was suffering from the ravages of sin.

### *The Kingdom of God as Restoration*

Is this a peace that will come in the normal course of events? Have we in human history experienced a gradual movement toward the harmony and wholeness of shalom? May we say that the Kingdom is closer to its complete realization than it has ever been? A thoughtful consideration of our current social and ecological situations tells us that much of human activity is pushing things in exactly the opposite direction. Instead of restoring creation, our society is enslaving it—polluting, wasting resources, extinguishing species, degrading ecosystems. Instead of bringing justice and harmony to human relationships—surely shalom means this—our world system perpetuates deep poverty and wasteful riches.

The modern worldview does not apparently lead us in the direction of shalom. But should we be surprised at this? After all, there is a rebellion going on. Another kingdom is striving for dominance in world affairs, and God's shalom is not on its agenda. Human sinfulness is quite thoroughly expressed in the affairs of people and nations. In particular, our devotion to material prosperity as our highest good has led to human and environmental brokenness of global proportions.

A passage in Romans speaks to this situation: "The creation waits in eager expectation for the sons of God to be revealed

. . . the creation itself will be liberated from its bondage to decay and brought into the glorious freedom of the children of God" (Rom. 8:19, 21). This liberation is another expression of shalom, the peace of wholeness. So we see a restoration to wholeness as the outcome of Christ's redeeming work, and it is a restoration that involves those who are redeemed. Jesus said, "I tell you the truth, anyone who has faith in me will do what I have been doing. He will do even greater things than these, because I am going to the Father" (John 14:12). Wolterstorff puts it well:

> Can the conclusion be avoided that not only is shalom God's cause in the world but that all who believe in Jesus will, with him, engage in the works of shalom? Shalom is both God's cause in the world and our human calling. Even though the full incursion of shalom into our history will be divine gift and not merely human achievement, even though its episodic incursion into our lives now also has a dimension of divine gift, nonetheless it is shalom that we are to work and struggle for. We are not to stand around, hands folded, waiting for shalom to arrive. We are workers in God's cause, his peace-workers.

This, in brief, is the task of stewardship.

## Justice and Peace: Reforming the Culture

Humankind, created in God's image, was given by God the cultural mandate: the capacity and the mandate to work toward the building of a culture. Human history records the progress that we have made in fulfilling this aspect of dominion—the accomplishments of art, music, science, technology, and the construction of cities. These are representative of a culture developed to the point of great mastery over the natural world and the social world. The potentials present in the created order have been developed in imaginative and diverse fashion, and it is fair to say that much of this is worthy of praise and amazement. But not all that we have done is good; since culture means not only the things that we have made but also the structures and arrangements of our social order, it is clear that sinfulness permeates our culture.

Wolterstorff speaks of this world of ours as "a world of deep sorrows," and documents several categories of sorrows—largely brought on by the social system that we have created. There are sorrows of injustice—the economic and political arrangements that condemn millions to deep poverty or the loss of personal freedom. And there are sorrows of "misplaced values," the result of the diversion of great wealth to the building of armaments or the enjoyment of luxuries. There are also sorrows of undesired consequences—miseries that arise from the workings of our social order—the loss of traditions, of a sense of belonging, the boredom of meaningless work, and so forth. To this list we must add the serious threats to the social order brought on by the limits of natural resources and pollution of earth, air, and water.

"This," says Wolterstorff, "is the world we have made for ourselves. It is a world with which only the privileged, and the imperceptive at that, could be satisfied. What we must now begin to consider is how Christians should act in this world, and how, as they see it, others should act as well. In a world such as this, what should be our project?"

The "project" Wolterstorff refers to is struggle for the reform of the culture that our society has developed. This is one of the great mandates of the Kingdom of God (there are others, also of great importance—spreading the good news of salvation, for example, or as we shall soon see, redeeming the creation).

This reforming work implies that there is a standard with which we can compare the structures and practices of our culture, a norm that tells us how things should be. Indeed, many of the workings of human sinfulness are obviously a distortion of that which is good, and such wrongdoing is broadly recognized as evil, immoral, unethical, or illegal. Misuse of the political process—as in graft and conspiracy, deliberate violations of the rights of individuals, the immoral practices surrounding sexuality—these are clearly in violation of Christian morality, if not societal morality. It is a fairly straightforward matter to find biblical grounds for judging such behavior and for demonstrating the direction of renewal and reform. Beyond these personal wrongs, however, is the class of wrongdoing known as *structural evil*, the remorseless workings of political, economic, and social structures and arrangements that result in injustice. What norm shall we use for judging our social order?

It is here that the biblical worldview has a vision to recommend to all of society: The light to which all of our culture must be held is the light of *shalom—God's peaceful order*. To embark on our project, then, is to recognize that having been reconciled to God, we are to become the ministers of reconciliation to our culture. We are called to build the peaceful community, to demonstrate what the Kingdom of God is like and to bring about the realization of that Kingdom in history. Where there is goodness, we should affirm it—to God's glory; where there is error, we should call attention to it and work toward its correction; where there is evil we must condemn it.

Wolterstorff argues convincingly that this task must begin by considering the economic dimension of our modern world system, for injustice is most profoundly imbedded here. Surely, he argues, shalom means the enjoyment of justice, of each person being guaranteed his or her rights. To dwell in shalom also means to be at peace in all relationships—with God, with one's fellow human beings, and with nature—and to enjoy or delight in those relationships. In this light, Wolterstorff holds up one dominating issue that cries out for reform: *poverty*. After documenting the extent and effects of poverty, he states that

> it is not the sheer fact of massive world poverty that is scandal to the church and all humanity; the scandal lies in the fact that this abject poverty is today not an unavoidable feature of our human situation,

and even more so in the fact that the impoverished coexist in our world-system with an equal number who live in unprecedented affluence. Poverty amidst plenty with the gap becoming greater: this is the scandal.

The pursuit of justice, then, is implicit in the meaning of shalom—God's peace. Justice does not rule when the sustenance rights of people are denied—their rights to food, clothing, shelter, healthy air, and water. In his book *Until Justice and Peace Embrace*, Wolterstorff documents the unmistakable biblical evidence that God is on the side of the poor, and goes on to show that Third World poverty is to a great extent the result of economic domination by the industrialized nations. Stewardship—that part of the task that is directed toward reforming the culture—requires that God's people position themselves with God on the side of the poor. We who would show what the Kingdom of God is like need to be working for the kinds of changes that would bring justice and peace to the social order.

Certainly there is much more to this task of reforming the culture. All human culture, as Wolters points out in *Creation Regained*, demonstrates goodness, the evidence of God's creational norms—as well as the perversions that are a result of the sinful expressions of culture. Our hard task is to bring renewal and health to the culture even as we point out that which is not worth preserving.

### Wholeness in the Natural Order: Redeeming the Creation

Dominion over creation is the delegated responsibility to rule as God's image-bearers. Creation continues to belong to its Creator, and the concept of stewardship captures well the proper relationship of humankind to the rest of the natural order. We considered in chapter 9 the implications of this task, and concluded that as God's stewards we are responsible for the welfare of creation—we are to serve and preserve it—and love it—even while we make use of it for our culture-building purposes. Yet as we saw in chapter 12, we have been poor stewards—indeed, foolish stewards—in light of the deterioration of the natural ecosystems that support our life and economic activity. We focused on four dynamics that are largely responsible for this: population growth, pollution, resource limits, and biological depletion. We considered the claims of justice, and suggested that justice must be applied both to humanity and to the land and creatures.

We have seen that God's standard for justice is the standard of shalom—of wholeness and peace. This applies as much to the rest of creation as it does to human relationships. It follows, then, that to demonstrate what God's Kingdom is like is to bring shalom—to restore integrity—to those natural systems that have suffered at our hands. Here, then, is another project for Christian action in this world.

This is not simply a matter of bringing cosmetic surgery to our parks, playgrounds, and roadsides, planting shrubs and picking up the trash. Entire biomes (major plant and animal groupings) are suffering from misuse: Deserts are spreading over tropical savannahs as a result of overgrazing and wood gathering; tropical rain forests are being cut down to make way for grazing cattle or raising cash crops; species of plants and animals are being extinguished as a result of our destruction of their habitat; the global climate is being altered as a result of the burning of fossel fuel. From local ecosystems to the global biosphere, natural systems are being manipulated, degraded, and destroyed. How ever are we to bring healing when the problems are so extensive, so complex? In other words, how is shalom to be translated into a program for action?

To carry out this task, we need normative information—we must have firm knowledge of the workings of natural ecosystems and the ways that human activities interact with those systems (ecological and environmental science). We know that ecosystems have been around for a long time, and that they have an amazing resiliancy, a capacity to absorb wastes and to yield some harvest. Surely it is equivalent to shalom—to integrity and wholeness—when we participate in these systems only to the extent that the natural systems can continue to exist in time and space. This is the only pathway to a sustainable future. Such knowledge is being gathered and evaluated on a regular basis through the work of organizations like the Worldwatch Institute. It is probably fair to say that most of the essential normative information is already in our grasp. Yet the human population continues to expand, the pollutants continue to be poured out on the land and into the air. What, might we ask, is missing?

The missing quality is *stewardship*. God's normative truth—the reality of stewardship of the creation—is a vital element of the biblical worldview, even though the message seems to be slow to permeate the Christian church. The modern worldview lacks this vision of humankind as steward over the rest of the natural order—and it has never been more needed. Without stewardship, it is inevitable that much of nature will perish, carrying with it first the humanity that is tied most closely to the land—the poor and hungry in the Third World.

What is needed is the most radical kind of stewardship—the kind that elevates stewardly action over political and economic concerns. The global world economy, based as it is on profitmaking and not—most definitely not—on justice or stewardship, has run the show for too long. Economist Bob Goudzwaard has called the ideology of economic growth one of the dominant gods of our time—we have worshiped at its feet, and it has betrayed us. Listen to Lester Brown and Edward Wolf in *State of the World* 1987:

> No widely shared vision exists of the need for worldwide progress to stabilize population, control carbon emissions, and revolutionize energy-using technologies. No agenda or five-point plan has been

drafted that confronts or even acknowledges the most serious challenges facing the world in the decades ahead.

This must be stewardship at all levels—from the backyard to the biosphere. Change is coming—it will be imposed on us by the environment, or else we will be wise and adjust our interactions with the environment before that happens. Those people and nations that embrace stewardship as their priority will be following the path of God's wisdom, whether they acknowledge him or not.

What of the United States? Where will we align ourselves? Will we make it our number one priority to seek dominance of the world economy, to the detriment of the poor nations as well as the global environment? Will we, for example, continue to approve the manipulations of multinational corporations that exploit the poor in Third World nations? Will we divert our energy use toward coal, recognizing that it is abundant and will soon be economically superior to oil? If we do, we guarantee that the greenhouse effect will be on us all the sooner. Will we continue to devote the lion's share of our national budget on defense? Will we be content to divert a diminishing percentage of our gross national product to aid for the developing world?

The global dimensions of the environmental problems require that nations begin to act in concert in taking positive action toward the goal of a sustainable earth. The consequences of national policies extend beyond national borders, therefore the policies themselves must be seen in their global perspective. An outstanding example of this principle (and an encouraging sign that the United States is offering stewardly leadership in some issues) can be seen in the Montreal Protocol of September 1987. Twenty-four nations negotiated a treaty calling for a reduction in world production and use of chlorofluorocarbons (CFCs), chemicals that are linked to the destruction of the stratospheric ozone layer. At this meeting, the U.S. Environmental Protection Agency took the lead in proposing stiff restrictions that were opposed by the Common Market nations, Japan, and the USSR. The final treaty was a compromise that calls for cutting CFC consumption 50 percent by 1999.

Brown and Wolf refer to *Centers of Decision*, small groups of nations whose role is especially critical for one or another of the globally significant problems. Coal use, for example, in the United States, China, and the Soviet Union will play a dominant role in the speed of advancement of the greenhouse warming and sea level rise that are tied to increasing carbon dioxide in the atmosphere. Brazil, Indonesia, and Zaire contain half of the world's remaining tropical forests; they hold the key to the future of the forests and to the certain impact on global climate if those forests are removed. Hear the remarkable conclusions of Brown and Wolf:

> A sustainable future calls upon us simultaneously to arrest the carbon dioxide buildup, protect the ozone layer, restore forests and soils, stop population growth, boost energy efficiency, and develop renewable energy sources. *No generation has ever faced such a complex set of issues requiring immediate attention . . . The course corrections needed to restore a worldwide improvement in the human condition have no precedent*. And they may not be possible if the militarization that is hampering international cooperation and preempting leadership time, fiscal resources, and scientific personnel continues. Anyone contemplating the scale of the needed adjustments is drawn inescapably to one principle conclusion: The time has come to make peace with each other so *that we can make peace with the earth*.

### The Reforming Project

The Kingdom of God will come. God has acted in history to redeem his creation, and although that redemption will not be complete until the return of Christ, we as God's people have the responsibility to act as God's agents in demonstrating what the Kingdom is like and, to the extent possible, bringing the

Kingdom to reality in this world. Our work is a reforming task; the Fall has corrupted both human culture and the created world, and both are in need of the healing word of shalom, the restoration of wholeness. It seems a forbidding task as we consider the multitude of problems and the deep penetration of sinfulness into the very structure of human affairs. Where can be begin?

As we saw in chapter 12, we must first learn what needs to be done (the knowledge base), and we must show that it is right to do so (the ethical base). But we must then move to *the most crucial step—to take action*, to go beyond developing a position or understanding what our Christian duty is, and do what needs to be done. One dimension to the reforming project that can involve all of us is the dimension that involves personal lifestyles. We can learn what it means to live responsibly and then do it. We also can promote stewardship in our families, in our churches, in the institutions that employ us.

Another dimension of stewardship involves developing a position on larger issues, and here I recommend the paradigm developed in chapter 12. The point of this process is to extend your influence by doing the scholarly and the practical work that is directed toward resolving issues. Perhaps the most difficult dimension to stewardship is the extension to major corporate and political structures—reaching even to the international scene. For most of us, our participation in the political process as voters represents the only stewardly action we can take to influence large-scale issues like acid rain or the global carbon cycle. Some, however, will have the opportunity to participate directly in such issues by working for the Environmental Protection Agency, or for development-minded Christian missions, or for the Peace Corps. Surely all of these are high callings!

I will refrain from the temptation to present the details of stewardly action. It would be incomplete, and it would not be appropriate for this supplementary text. Others have attempted this—for example, the authors of *Earthkeeping* offer some thirty stewardship principles for consideration, addressing actions on all levels. If you are a college student, you could do no better than to attend a summer session at Au Sable Institute in the Michigan north woods. The mission of this organization is to promote Christian stewardship through a variety of programs, one of which offers courses for participating Christian colleges. Here you will learn what it means to be caretakers of land and water resources, and you may have opportunity to become involved with Au Sable's educational services to the local schools as an intern.

### *Reforming Biology*

Finally, what of the task of reforming biology? This is a responsibility facing those of us who are Christian biologists. We must seek the deepest possible understanding and penetration into the accumulated knowledge and procedures of the life sciences. As we do this, we must avoid the quick and easy judgment. This knowledge is largely correct; what has been learned is primarily the result of honest and painstaking scholarship on the part of thousands of people whose basic goal has been to discover the truth. We should affirm this work, recognizing that it has uncovered God's creational activity. We should not be so naive, however, that we do not expect to see the evidence of human sinfulness reflected in the knowledge and practice of our science. Reforming biology, then, means that we should look for this evidence as we read the works that seek to interpret the life sciences, for here is where worldviews are especially significant.

In particular, our reforming task will be tested as we examine those issues that have most strongly influenced human affairs—the four revolutions I have presented in this text. The Darwinian revolution continues to generate great tensions because of its pronouncements on origins. The biomedical revolution challenges Christian ethics by its direct effects on our reproduction, our health and length of life. The genetic revolution promises to unravel the very instructions that describe living organisms, and challenges us to respond to the potential for changing those instructions. And the environmental revolution calls on us to help to shape the future, to be involved in the Kingdom of God as a place on earth where justice and peace reign.

It is our privilege as biologists to study God's creation. Because we understand it in greater depth than others, we are also more responsible for its use and its integrity. Because we worship the One who created life, we are able to take the works of our mind and our hands and offer them to him as suitable objects for his Lordship. And because we can see God's wisdom and beauty in living things, we can join them in bringing him the glory and praise that he deserves: "*Let everything that has breath praise the Lord*" (Psalm 150:6).

## Summary

The theme of worldviews is once again brought into focus. *Naturalism* and *theism* are major opposing worldviews with implications for all of life. In scientific work, worldviews are shaping principles that are rarely explicit but often play an important role in issues involving science and faith. *Evolutionism*—a worldview extension of evolution—strongly influences thinking on issues involving origins. *Secularism* and *materialism* are two well-entrenched worldviews with great influence on our culture's approach to resources and the environment. These have become the dominant components of the modern worldview.

Yet it seems clear that the modern worldview has not served us well. We cannot be content with the way things are going, yet we do not seem to know how to proceed—nor do other cultures with other worldviews. On the other hand, the Christian worldview brings a vision for how life should be, and an agenda for realizing that vision. We are enlisted in the tasks of proclaiming the gospel, of reforming the culture, and of redeeming the creation.

The *Kingdom of God* appears as a unifying strand through the Bible, referring to a time when God will reign over the cosmos and peace and wholeness will prevail. That Kingdom—inaugurated by Jesus Christ—is in progress, although it is being contested by the forces of evil. The major theme of the Kingdom is *shalom*, or peace—referring to the harmony and wholeness that will be restored to the creation as a result of the reconciling work of Christ.

But we are not there yet. God calls us to participate in the works of shalom, of redemption, and this is the task of stewardship. One element of this task is *reforming the culture*. It is permeated with sin, and our calling is to hold all parts of culture to the standard of God's shalom. We must affirm goodness, correct error, and condemn evil in our culture. In particular, we should position ourselves with God on the side of the poor, since shalom means bringing justice to human relationships.

The other major element of our stewardship task is *redeeming the creation*. The standard of shalom—wholeness—applies as well to the natural world as to human relationships. This means bringing healing to a creation that shows extensive damage from our mismanagement. This task is based on knowledge of how natural systems work, and how we interact with them. Although much of this knowledge is in place, modern society does not seem ready to act accordingly.

The corrective vision of Christian stewardship over the natural order is missing from this picture. This stewardship at all levels of human interaction with the environment must take preeminence over economic concerns. The global dimensions of this responsibility are illustrated by an international treaty for the reduction of those chemicals responsible for destroying the ozone layer. Here is a clear example of the kind of corrective action that must become applied to many areas of concern.

The Kingdom of God will come. The reforming task facing us demands that we move from knowledge to action as we undertake to do our Christian duty. From personal lifestyles to our citizenship duties to involvement with governmental agencies, the call to stewardly action must be given.

The specific task of reforming biology is assigned to Christians in the life sciences. We must affirm what is correct, but should also expect to find human sinfulness reflected in the practice and expressions of our science. Our reforming task will especially be challenged by the issues we have examined in the four biological revolutions—Darwinian, biomedical, genetic, and environmental. Finally, the perspective of biology as the study of God's creation gives us a special privilege and a special responsibility to care about this area of knowledge and bring glory to God.

# NOTES

## 1. Biology and Worldviews

1. "You know, our" William Cronon, *Changes in the Land: Indians, Colonists and the Ecology of New England* (New York: Hill and Wang, 1983), 162.
9. "a set of" James W. Sire, *The Universe Next Door: A Basic World View Catalog* (Downer's Grove, IL: InterVarsity Press, 1976), 17.
9. "the comprehensive framework" Albert M. Wolters, *Creation Regained: Biblical Basics for a Reformational Worldview* (Grand Rapids, MI: Eerdmans, 1985), 2.
10. I am indebted to Brian J. Walsh and J. Richard Middleton, *The Transforming Vision: Shaping a Christian World View* (Downer's Grove, IL: InterVarsity Press, 1984), for their insight into the crucial role of worldviews in matters that relate to faith and knowledge.
12. "God the Father" Herman Bavinck, from Wolters, *Creation Regained*, 10.
13. "Cosmos is all that" Carl Sagon, *Cosmos* (New York: Random House, 1980), 4.

## 2. God and His World

16. The distinction between nature and creation is from Wesley Granberg-Michaelson, *A Worldly Spirituality* (San Francisco: Harper & Row, 1984), 53 ff.
18. ff. Reference to Creation by Word and Wisdom from Brian J. Walsh and J. Richard Middleton, *The Transforming Vision: Shaping a Christian World View* (Downer's Grove, IL: InterVarsity Press, 1984), 44 ff.
20. "the origin of" George Gaylord Simpson, *The Meaning of Evolution*, rev. ed. (New Haven, CT: Yale University Press, 1967), 279.
21. I am indebted to Uko Zylstra of Calvin College (personal communication) and to A. L. Wolters, *Creation Regained: Biblical Basics for a Reformational Worldview* (Grand Rapids, MI: Eerdmans, 1985), 15 ff., for insights into Creational Law.
24. On biological death, see F. Van Dyke, *American Science Affiliation* 38 (1986):11 for a recent paper on this view (Theological Problems of Theistic Evolution).

**25. ff.** My thanks to Cal DeWitt (personal communication) for ideas expressed here.

**25.** The many passages from Psalms include Psalms 148; 96; 97:6; 98:7–9; 100:1–3; 103:22; 145:10, 21; 146:1, 2; 149:1, and 150:6.

## 3. The Scientific Enterprise

**31.** "the study of" Ernst Mayr, *The Growth of Biological Thought* (Cambridge: Harvard University Press, 1982), 835.

**32–34.** This work is published in R. T. Wright, R. B. Coffin, C. P. Ersing, and D. Pearson, "Field and Laboratory Measurements of Bivalve Filtration of Natural Marine Bacterioplankton," *Limnology and Oceanography* 27 (1982): 91–98.

**36. ff.** I am deeply indebted to Del Ratzsch, *Philosophy of Science: The Natural Sciences in Christian Perspective* (Downer's Grove, IL: InterVarsity Press, 1986), for his insights into the structure of science. His book has greatly contributed to the thoughts of this chapter.

**38. ff.** Harry Cook (personal communication) has contributed greatly to the ideas expressed here.

**40.** "the Kuhnian movement" Ratzsch, *Philosophy of Science*, 55.

**40.** Thanks to Loren Wilkinson, "New Age, New Consciousness, and the New Creation," in Wesley Granberg-Michaelson, ed., *Tending the Garden: Essays on the Gospel and the Earth* (Grand Rapids, MI: Eerdmans, 1987), 6–29, for his insights into New Age thinking.

**41.** "a powerful alternative" Loren Wilkinson, "New Age, New Consciousness, and the New Creation," in W. Granberg–Michaelson, ED., *Tending the Garden* (Grand Rapids, MI); Eerdmans, 1987), 27.

**41–44.** I am indebted to Ernst Mayr, *Growth of Biological Thought*, 43 ff., for these insights into the structure of biology.

**44.** Common Sense Approach to Science is Adopted from David Price, John L. Wiester, and Walter R. Hearn, *Teaching Science in a Climate of Controversy* (Ipswich, MA: American Scientific Affiliation, 1986), 23.

**45. ff.** Thanks to Uko Zylstra (personal communication) for insights into the limitations of science.

## 4. Relating Science and Christianity

**51.** "By every criterion" John C. Greene, "The Kuhnian Paradigm and the Darwinian Revolution in Natural History," in D. Roller, ed., *Perspectives in the History of Science and Technology* (Norman, OK: University of Oklahoma Press, 1971), 5.

**52.** I am indebted to Ian Barbour, *Issues in Science and Religion* (Englewood Cliffs, NJ: Prentice-Hall, 1966), 37–40, for these insights into natural theology.

**52.** "By the Oeconomy," Carl Linnaeus, in Donald Worster, *Nature's Economy: A History of Ecological Ideas* (Cambridge, U.K.: Cambridge University Press, 1977), 37.

**52.** Ernst Mayr, *The Growth of Biological Thought* (Cambridge: Harvard University Press, 1982), 104, clearly indicates this distinction between the physical scientists and the natural historians.

54. "the works created" John Ray, in John C. Greene, *Darwin and the Modern World View* (Baton Rouge: Louisiana State University Press, 1961), 40.
54. ff. Mayr, *Growth of Biological Thought*, chapters 4 and 7, is my source for this information on the natural historians.
55. "Nature, in successively" Jean Baptiste Lamarck, in Ernst Mayr, *The Growth of Biological Thought* (Cambridge: Harvard University Press, 1982), 353.
56. "The old argument" Charles Darwin, in Greene, *Darwin and the Modern World View*, 43.
57. "The conclusion of" Charles Hodge, in Fred Gregory, "The Impact of Darwinian Evolution on Protestant Theology in the Nineteenth Century," in David C. Lindberg and Ronald L. Numbers, *God and Nature: Historical Essays on the Encounter Between Christianity and Science* (Berkeley: University of California Press, 1986), 375 ff.
58. See Mayr, *Growth of Biological Thought*, 105, for this acknowledgment.
59. Recent views on warfare between science and Christianity from Lindberg and Numbers, *God and Nature*, 8 ff.
60. "Let each man" Charles Darwin, as recorded in John C. Greene, *Darwin and the Modern World View* (Baton Rouge, LA: Louisiana State University Press, 1961), 45.
60. "Agnosticism, in fact" Thomas H. Huxley, and views on Darwin and Huxley, from A. Hunter Dupree, "Christianity and the Scientific Community in the Age of Darwin," in Lindberg and Numbers, *God and Nature*, 351 ff.
61. "Understand that all" Thomas H. Huxley, Ibid., 365.
62. "The scientist's devotion" E. O. Wilson, in Conrad Hyers, *The Meaning of Creation* (Atlanta: John Knox Press, 1984), 17.
67. Robert B. Fischer, *God Did It, But How?* (Grand Rapids, MI: Zondervan, 1981), 95.

## 5. Perspectives on Genesis 1

71. "Nothing is here" John Calvin, *Commentaries on the First Book of Moses Called Genesis*, Vol 1, John King, translator (Grand Rapids, MI: Eerdmans, 1948), 84.
73. "Is it ever" E. C. Lucas, "Some Scientific Issues Related to the Understanding of Genesis 1–3," *Thermelios* 12 (1987): 46–51.
74. "The word 'hymn'" Henri Blocher, *In the Beginning* (Downer's Grove, IL: InterVarsity Press, 1984), 32. My debt to Blocher goes far beyond the several quotes included in this chapter.
74. "Beyond any doubt" *Ibid.*, 34.
75. "In the act" *Ibid.*, 26.
77. Dallas Cain, "Let There Be Light: Spectrum of Creation Theories," *Eternity* (May 1982): 20, 21.
77. "It is by" Footnote in *Scofield Reference Bible*. (New York, NY: Oxford Univ. Press, 1917), 4.
78. John Wiester's work is from his book *The Genesis Connection* (Nashville: Thomas Nelson Publishers, 1983), chapter 13.

**78.** "Scientists are now" *Ibid.*, 209.
**79.** "How did Moses" *Ibid.*, 210.
**80.** "The Bible is" Creation Research Society, cited in Peter Zettenberg, ed., *Evolution and Public Education* (St. Paul, MN: University of Minnesota, 1981), 37.
**81.** "The rejection of" Blocher, *In the Beginning*, 48.
**82. ff.** The Framework theory is described by Blocher, *Ibid.*, 49 ff.; Conrad Hyers, *The Meaning of Creation* (Atlanta: John Knox Press, 1984), 67 ff,; and James M. Houston, *I Believe in the Creator* (Grand Rapids, MI: Eerdmans, 1980), 60, 61; and appears in many other works.
**84.** "What are we" Blocher, *In the Beginning*, 59.
**84.** "While Blocher's framework" Pattle P. T. Pun, "A Theology of Progressive Creationism," *Perspectives on Science and Creation Faith* 39: 15.
**87.** "In order to" Blocher, *In the Beginning*, 23.
**89.** "It is not" Del Ratzsch, *Philosophy of Science: The Natural Sciences in Christian Perspective* (Downer's Grove, IL: InterVarsity Press, 1986), 132, 133.
**89–90.** See Charles E. Hummel, *The Galileo Connection* (Downer's Grove, IL: InterVarsity Press, 1986), and Howard J. Van Till, *The Fourth Day* (Grand Rapids, MI: Eerdmans, 1986) for two recent presentations of complementarism.

## 6. The Origin of Life

**95.** "And, therefore, gentlemen" Louis Pasteur, as reported in René Dubos, *Pasteur and Modern Science* (Garden City, NY: Doubleday, 1960), 60.
**97.** "We tell this" George Wald, "The Origin of Life," *Scientific American* (August 1954): 47.
**97.** "The first theory" Robert Jastrow, *Until the Sun Dies* (New York: Warner Books, 1977), 51–52.
**98.** Table 2 is taken from C. G. Thaxton, W. L. Bradley, and R. L. Olsen, *The Mystery of Life's Origin: Reassessing Current Theories* (New York: Philosophical Library, 1984), 15. I owe a deep debt to these authors for their careful work, from which I have taken much of my discussion of origin-of-life issues.
**100.** "With further cooling" Sylvia S. Mader, *Biology, Evolution, Diversity and the Environment* (Dubuque, IA: W. C. Brown, 1985), 254.
**100.** "a number of" K. Norstog and A. J. Meyerriecks, *Biology*, 2d ed. (Columbus, OH: C. E. Merrill, 1985), 40.
**100.** "a combination of" Douglas J. Futuyma, *Science on Trial: The Case for Evolution* (New York: Pantheon Books, 1983), 95.
**100.** It's likely that" *Ibid.*, 96.
**100.** "It can be" Michael Ruse, *Darwinism Defended* (Reading, MA: Addison Wesley, 1982), 168.
**100.** "As we move" *Ibid.*, 162.
**100.** "The origin of" Francis Crick, in Michael Pitman, *Adam and Evolution* (London: Rider Press, 1984), 148.
**101.** "too much speculation" Francis Crick, in Thaxton *et al.*, *Mystery of Life's Origin*, 195.

**101.** The reference to Fox and Dose, and the reference to circular reasoning, are in *Ibid.*, 77.
**102.** "In general, we" Erich Dimroth and Michael Kimberley, as quoted in *Ibid.*, 93.
**102.** "These experiments . . . claim" J. Brooks and G. Shaw, as quoted in *Ibid.*, 110.
**103.** "Curiously, the findings" Ernst Mayr, *The Growth of Biological Thought* (Cambridge: Harvard University Press, 1982), 583.
**103.** "The macromolecule-to-cell" D. G. Green and R. F. Goldberger, quoted in Thaxton, *et al.*, *Mystery of Life's Origin*, 179.
**106.** Details of current views on proof and falsification can be found in Del Ratzsch. *Philosophy of Science: The Natural Sciences in Christian Perspective* (Downer's Grove, IL: InterVarsity Press, 1986), chapters 5 and 6.
**106.** Information on observational-comparative method from Mayr, *Growth of Biological Thought*, 30–32.
**108.** "No intelligent Creator" Pitman, *Adam and Evolution*, 255.
**111.** "Anyone who is" David Wilcox, "A Taxonomy of Creation," *Journal of American Science Affiliation* 38 (1986): 244–250, this quote page 250.

## 7. The Darwinian Revolution

**115.** "The Darwinian revolution" Ernst Mayr, *The Growth of Biological Thought* (Cambridge: Harvard University Press, 1982), 501.
**116. ff.** *Ibid.*, 505 ff.
**116.** "For many biologists" *Ibid.*, 507.
**119.** Books critiquing evolution are listed in Further Reading for this chapter; see Denton (1985), Pitman (1984), and Rifkin (1983).
**120.** "The claim—or" Mayr, *Growth of Biological Thought*, 438.
**120.** "By providing a" *Ibid.*, 510.
**121.** "It seems to" Charles Darwin, as recorded in A. Hunter Dupree, *Christianity and the Scientific Community in the Age of Darwin*, in David C. Lindberg and R. L. Numbers, eds., *God and Nature: Historical Essays on the Encounter Between Christianity and Science* (Berkeley: University of California Press, 1986), 351–368.
**122.** "The most closely" Alfred Russel Wallace quoted in Mayr, *Growth of Biological Thought*, 419.
**126.** Details on the Neo-Darwinian synthesis from *Ibid.*, 566 ff.
**127.** A discussion of punctuated equilibrium and other recent changes in the synthetic theory is found in G. L. Stebbins and F. J. Ayala, "The Evolution of Darwinism," *Scientific American* (July 1985): 72–82.
**129.** "It is our" P. S. Moorhead and M. M. Kaplan, eds., *Mathematical Challenges to the Neo-Darwinian Interpretation of Evolution* (Philadelphia: Wister Institute Press, 1967), 221.
**129.** The evolutionist response to the mathematicians is found in Mayr, *Growth of Biological Thought*, 429.
**131.** Discussion on tautology and natural selection in Jeremy Rifkin, *Algeny* (New York: Viking Press, 1983), 136 ff.

**133.** Table 3 is from Robert B. Fischer, *God Did It, But How?* (Grand Rapids, MI: Zondervan, 1981), 64. In discussing this table, I am deeply indebted to Fischer's ideas

**134.** "The theory of" Duane Gish, "A Consistent Biblical and Scientific View of Origins," Derek Burke, ed., *Creation and Evolution: When Christians Disagree* (Leicester, UK: InterVarsity, 1985), 140, 141.

**135.** "Since most creationists" Davis Young, *Christianity and the Age of the Earth* (Grand Rapids, MI: Zondervan, 1982), 149.

**135.** "I find it" *Ibid.*, 151.

**136.** "I believe Creationism" Michael Ruse, *Darwinism Defended* (Reading, MA: Addison-Wesley, 1982), 303.

**137.** "To my mind" Charles Darwin, *The Origin of Species* (1859), The Harvard Classics, vol. 11 (New York: P. F. Collier and Son, 1909), 527–529.

## 8. Where Are You, Adam?

**142.** "A . . . cry of" John C. Greene, *The Death of Adam* (Ames, IA: The Iowa State University Press, 1959), 14.

**143.** See S. J. Gould, "Knight Takes Bishop?" *Natural History* 95 (1986): 18–33.

**144.** "a superior intelligence" Alfred Russel Wallace, as cited in Michael Ruse, *Darwinism Defended* (Reading, MA: Addison-Wesley, 1982), 234.

**145. ff** I am indebted to Dick Hodgson of Dordt College for allowing me to consult his Historical Geology supplementary papers for data on fossil hominids and much of the basis for Table 4.

**145.** Gould anecdote from Stephen Jay Gould, *Ever Since Darwin: Reflections in Natural History* (New York: American Museum of Natural History, 1977), 56.

**145.** "missing links" *Science and Creationism: A View from the National Academy of Sciences* (Washington, D. C.: National Academy Press, 1984), 23.

**145.** "a succession of" *Ibid.*, 24.

**147.** Mayr reference from Ernst Mayr, *The Growth of Biological Thought* (Cambridge: Harvard University Press, 1982), 621.

**147.** "mosaic evolution" *Ibid.*, 622.

**148.** "The paramountly important" Douglas J. Futuyma, *Science on Trial: The Case for Evolution* (New York: Pantheon Books, 1983), 111.

**148.** "Just what the" *Ibid.*, 112.

**148.** An excellent critique of sociobiology from a Christian perspective by Paul and Mary Ellen Rothrock (Taylor University) is found in *Perspectives on Science and Christian Faith* 39 (1987): 87–93.

**149.** "Humans are more" Richard Leakey, *The Making of Mankind* (New York: E. P. Dutton, 1981), 20.

**149.** "Who is to" Greene, *The Death of Adam*, 332.

**151.** "The style is" Henri Blocher, *In the Beginning* (Downer's Grove, IL: InterVarsity Press, 1984), 34–35.

**152.** "The presence of" *Ibid.*, 155–156.

**152.** "leaving before cleaving" Derek Kidner, *Genesis: An Introduction and Commentary* (London: InverVarsity Press, 1967), 66.
**153.** "to be the" Blocher, *In the Beginning*, 85.
**154.** Houston reference from James, M. Houston, *I Believe in the Creator* (Grand Rapids, MI: Eerdmans, 1980), 77 ff.
**155.** "Two sentences could" Blocher, *In the Beginning*, 137. Reference to his discussion of motives, 138 ff.
**156.** "Thus, it is" *Ibid.*, 140–141.
**158.** "It is perhaps" *Ibid.*, 231.

## 9. Stewards of Creation

**162.** "the rule of" Gandalf, in J. R. R. Tolkien, *The Return of the King* (New York: Ballantine, 1965), 33 ff.
**164.** "The great Western" Ian L. McHarg, *Design with Nature* (Garden City: Doubleday, 1969), 26.
**164.** See Arnold Toynbee, "The Religious Background of the Present Environmental Crisis," in D. and E. Springs, eds., *Ecology and Religion in History* (New York: Harper & Row, 1974).
**165.** Lynn White, Jr., "The Historical Roots of Our Ecological Crisis," *Science* 155 (1967): 1203–1206.
**165.** "We would seem" *Ibid.*, 1206.
**165.** Wesley Granberg-Michaelson, *A Worldly Spirituality* (San Francisco: Harper & Row, 1984), 32 ff.
**166.** "Nature is something" *Ibid.*, 38–39.
**167.** "the modern mindset" *Ibid.*, 43.
**167–170.** Reasoning developed here was first published in Richard T. Wright, "Responsibility for the Ecological Crisis," *Bioscience* 20 (1970): 851–853.
**168.** "A culture's publicized" Yi-Fu Tuan, "Our Treatment of the Environment in Ideal and Actuality," *American Scientist* 58 (1970): 244–249.
**170.** Reference to naming the animals from Henri Blocher, *In the Beginning* (Downer's Grove, IL: InterVarsity Press, 1984), 91.
**171.** I am indebted to Brian J. Walsh and J. Richard Middleton, *The Transforming Vision: Shaping a Christian World View* (Downer's Grove, IL: InterVarsity Press, 1984), 54 ff., for these insights into the cultural mandate.
**171.** "serving and preserving" Granberg-Micahaelson, *A Worldly Spirituality*, 65.
**172.** Thanks to Loren Wilkinson, *Earthkeeping: Christian Stewardship of Natural Resources* (Grand Rapids, MI: Eerdmans, 1980), for insights into the impact of the Fall on creation.
**173.** "From Abraham (who" *Ibid.*, 213.
**174.** "for the welfare" *Ibid.*, 233.
**175.** "personal salvation . . . matter" Granberg-Michaelson, *A Worldly Spirituality*, 91 ff.
**176.** "the power of" *Ibid.*, 98.
**178.** The Cyrus theme was suggested by Cal DeWitt, director of Au Sable Institute of Environmental Studies (personal communication).

## 10. The Biomedical Revolution

**183.** "If ethical principles" A. V. Hill, *The Ethical Dilemma of Science* (New York: The Rockefeller University Press, 1960), 82.

**184.** "All knowledge, not" *Ibid.*, 87.

**185.** "not only have" Hans Jonas, "Freedom of Scientific Inquiry and the Public Interest," *The Hastings Center Report* 6 (1976): 15.

**185.** "the current biomedical" D. Gareth Jones, *Brave New People*, rev. ed. (Grand Rapids, MI: Eerdmans, 1985), 7.

**186–187.** I am indebted to Leon Kass for much of this information, as he is quoted in Thomas A Shannon, ed., *Bioethics* (Mahwah, NJ: Paulist Press, 1980), 381.

**188.** "overmedicalization of life" Thanks to Jones, *Brave New People*, 30 ff., for information on the biomedical model.

**189.** "According to the" Aaron Wildavsky, in Shannon, *Bioethics*, 529.

**190. ff.** I am indebted to philosopher Arthur Holmes of Wheaton College for much of this discussion on ethical systems and Christian ethics. See Holmes, *Ethics: Approaching Moral Decisions* (Downer's Grove, IL: InverVarsity Press, 1984).

**191.** "identifies values to" *Ibid.*, 11.

**192.** This information on the three basic bioethical principles is from Carol Levine, ed., *Taking Sides: Clashing Views on Controversial Bio-Ethical Issues* (Guilford, CT: Dishkin Publishing Group, 1984), 8–10.

**195.** Gareth Jones reference from Jones, *Brave New People*, 64 ff.

**195.** "Quality control . . . recognizes" *Ibid.*, 49.

**198.** "IVF is legitimate" *Ibid.*, 116. Much of my coverage of this topic is from this section of his book.

**198.** Abortion Data from Center for Disease Control, as reported in *The Boston Globe*, Nov. 26, 1988.

**199.** "Abortion places upon" *Ibid.*, 150.

**200.** "The heresy of" *Ibid.*, xi and xvii.

**200.** See Mark Coppenger, *Bioethics: A Casebook* (Englewood Cliffs, NJ: Prentice-Hall, 1985), for an excellent casebook on bioethics that arose from such a Christian college course.

## 11. The Genetic Revolution

**205.** "There has hardly" Ernst Mayr, *The Growth of Biological Thought* (Cambridge: Harvard University Press, 1982), 825.

**206.** The DNA structure discovery is described in James Watson, *The Double Helix* (New York: New American Library, 1968), 126.

**206.** "It has not" *Ibid.*, 139.

**208.** "The total human" Walter Gilbert, "Proposal to Sequence the Human Genome Stirs Debate," *Science* 232 (1986): 1598.

**208.** "the complete understanding" Walter Gilbert, "Shifting Sentiments Over Sequencing the Human Genome," *Science* 233 (1986): 621.

**208.** Human genome project information from "Watson Will Head NIH Genome Office," *Science* 241 (1988): 1752.

**209.** "Another paradigm had" Liebe F. Cavalieri, *The Double-Edged Helix* (New York: Columbia University Press, 1981), 26.

**210.** See *Science* 236 (June 5, 1987) for numerous articles on the applications of recDNA technology.

**212.** "from rudimentary but" David Baltimore, "A Milestone for Biology," *Boston Globe* (March 2, 1985).

**212.** I am indebted to Marvin Rogul's article "Risk and Regulation in Biotechnology," in A. H. Teich, M. A. Levin, and J. H. Pace, *Biotechnology and the Environment* (Washington, D.C.: American Association for the Advancement of Science, 1985). This publication is the proceedings of the seminar held for EPA.

**213.** "adequate scientific knowledge" Discussion and quote about the National Academy of Sciences report on the release of genetically engineered organisms was in the *Boston Globe* (August 24, 1987).

**214.** Information on gene therapy guidelines from "Gene Therapy Guidelines Revised," *Science* 228 (1985): 561–562.

**214.** "Today, our biotechnical" Jeremy Rifkin, *Algeny* (New York, Viking Press, 1983), 244.

**215.** "The scientist who" Cavalieri, *Double-Edged Helix*, 135.

**216.** "consider the real" *Ibid.*, 157.

**216.** Information on theological resolution from Carol Levine, ed., *Taking Sides: Clashing Views on Controversial Bio-Ethical Issues* (Guilford, CT: Dishkin Publishing Group, 1984), 274 ff.

**217.** See Robert L. Herrmann, ed., *Making Whole Persons: Ethical Issues in Biology and Medicine* (*Journal of the American Scientific Association*) 1980, for an excellent group of articles.

**218.** "God calls all" David L. Wilcox, "A Christian Integrative Framework for Biology," *Christian Scholar's Review* 12 (1983): 339–348, This quote page 346.

**218.** "A conviction governing" Henry Stob, *Ethical Reflections* (Grand Rapids, MI: Eerdmans, 1978), 221–222.

## 12. The Environmental Revolution

**224.** "The French use" Lester R. Brown, *the Twenty-ninth Day* (New York: W. W. Norton, 1978), 1.

**227.** "Can the world" Anonymous, *The Global Possible: Resources, Development and the New Century* (Washington, D.C.: World Resources Institute, 1984), Foreword.

**227.** "a world that" *Ibid.*, 1.

**228.** I am indebted to Daniel D. Chiras, *Environmental Science: A Framework for Decision Making* (Menlo Park, CA: Benjamin/Cummings, 1985), for this view of environmental science.

**231.** "whatever the consequences" Gus Speth, *Global Energy Futures and the Carbon Dioxide Problem* (Council on Environmental Quality, 1981), 3.

**231.** See Sandra Postel, *Altering the Earth's Chemistry: Assessing the Risks*, Worldwatch Paper 71 (Washington, D. C. Worldwatch Institute, 1986).

**232.** "Our contemporary world" Lester R. Brown and Jodi L. Jacobson, *Our Demographically Divided World*, Worldwatch paper 74 (Washington, D.C.: Worldwatch Institute, 1986), 5.

**235.** "could be catastrophic" *Ibid.*, 46.

**237.** Biological depletion is competently presented in Edward C. Wolf, *On the Brink of Extinction: Conserving the Diversity of Life*, Worldwatch Paper 78 (Washington D.C.: Worldwatch Institute, 1987). I am indebted to this publication for much of the information appearing on pages 000–000. (NOTE TO ED: current ms pages 237–239.)
**238.** "In North America" *Ibid.*, 13.
**242.** "several billion years" Holmes Rolston III, *Philosophy Gone Wild: Essays in Environmental Ethics* (Buffalo, NY: Prometheus Books, 1986), 218. References to Rolston's work are from this book.
**243.** "With a careful" Susan Bratton, "A Fierce Green Fire Dying: Christian Land Ethics and Wild Nature," 35, A paper presented at the 1987 Au Sable Forum, Mancelona, Michigan.
**246.** "As we consider" Anonymous, *The Global Possible*, 3–4.

## 13. When Earth and Heaven Are One

**250.** The worldview crisis is discussed in Brian J. Walsh and J. Richard Middleton, *The Transforming Vision* (Downer's Grove, IL: InterVarsity Press, 1984), 142 ff.
**255.** For Kingdom as a unifying strand, see Howard Snyder, *A Kingdom Manifesto* (Downer's Grove, IL: InterVarsity Press, 1985), 12. I am deeply indebted to this work and to that of Walsh and Middleton, *The Transforming Vision*, for insights into the Kingdom of God.
**255.** "The Kingdom is" Snyder, *A Kingdom Manifesto*, 12, 13.
**257.** "The Old Testament" *Ibid.*, 19.
**258.** "That shoot of" Nicholas Wolterstorff, *Until Justice and Peace Embrace* (Grand Rapids, MI: Eerdmans, 1983), 71, 72.
**259.** "Can the conclusion" *Ibid.*, 72.
**260.** "a world of" *Ibid.*, 42.
**261.** "This . . . is the" *Ibid.*, 43.
**261.** Insights into the normative reforming task from Albert M. Wolters, *Creation Regained* (Grand Rapids, MI: Eerdmans, 1985), 78 ff., and Wolterstorff, *Justice and Peace*, 126 ff.
**262.** "it is not" Wolterstorff, *Justice and Peace*, 74.
**265.** Bob Goudzwaard's *Idols of Our Time* (Downer's Grove, IL: InterVarsity Press, 1984) is an insightful look at modern ideologies and worldviews.
**265.** "No widely shared" Lester Brown and Edward Wolf, *State of the World 1987* (New York: W. W. Norton, 1987), 212.
**266.** Centers of Decision discussed *Ibid.*, 209 ff.
**266.** "a sustainable future" *Ibid.*, 213.

# FURTHER READING


### 1. Biology and Worldviews

Cronon, William. *Changes in the Land: Indians, Colonists and the Ecology of New England*. New York: Hill and Wang, 1983.

Sire, James W. *The Universe Next Door: A Basic World View Catalog*. Downer's Grove, IL: InterVarsity Press, 1976.

Walsh, Brian J. and J. Richard Middleton. *The Transforming Vision: Shaping a Christian World View*. Downer's Grove, IL: InterVarsity Press, 1984.

Wolters, Albert M. *Creation Regained: Biblical Basics for a Reformational Worldview*. Grand Rapids, MI: Eerdmans, 1985

### 2. God and His World

Blocher, Henri. *In the Beginning*. Downer's Grove, IL: InterVarsity Press, 1984.

Granberg-Michaelson, Wesley. *A Worldly Spirituality*. San Francisco: Harper & Row, 1984

Houston, James M. *I Believe in the Creator*. Grand Rapids, MI: Eerdmans, 1980.

Hyers, Conrad. *The Meaning of Creation*. Atlanta, GA: John Knox Press, 1984.

Simpson, George Gaylord. *The Meaning of Evolution*, rev. ed. New Haven, CT: Yale University Press, 1967.

Van Til, Howard J. *The Fourth Day*. Grand Rapids, MI: Eerdmans, 1986.

Walsh, Brian J. and J. Richard Middleton. *The Transforming Vision: Shaping a Christian World View*. Downer's Grove, IL: InterVarsity Press, 1984.

### 3. The Scientific Enterprise

Bube, Richard H. *The Human Quest*. Waco, TX: Word Books, 1971.

Fischer, Robert B. *God Did It, But How?* Grand Rapids, MI: Zondervan Publishing House, 1981.

Granberg-Michaelson, Wesley, ed. *Tending the Garden: Essays on the Gospel and the Earth*. Grand Rapids, MI: Eerdmans, 1987.

Kuhn, Thomas S. *The Structure of Scientific Revolutions*. Chicago: University of Chicago Press, 1962.

Lovelock, James. *Gaia: A New Look at Life on Earth*. New York: Oxford University Press, 1979.

Mayr, Ernst. *The Growth of Biological Thought*. Cambridge: Harvard University Press, 1982.

Price, David, John L. Wiester, and Walter R. Hearn. *Teaching Science in a Climate of Controversy*. Ipswich, MA: American Scientific Affiliation, 1986.

Ratzsch, Del. *Philosophy of Science: The Natural Sciences in Christian Perspective*. Downer's Grove, IL: InterVarsity Press, 1986.

Wright, R. T., R. B. Coffin, C. P. Ersing, and D. Pearson. "Field and Laboratory Measurements of Bivalve Filtration of Natural Marine Bacterioplankton," *Limnology and Oceanography* 27 (1987): 91–98.

### 4. Relating Science and Christianity

Barbour, Ian. *Issues in Science and Religion*. Englewood Cliffs, NJ: Prentice-Hall, 1966.

Bube, Richard H., ed. *The Encounter Between Christianity and Science*. Grand Rapids, MI: Eerdmans, 1968.

Fischer, Robert B. *God Did It, But How?* Grand Rapids, MI: Zondervan, 1981.

Greene, John C. *Darwin and the Modern World View*. Louisiana State University Press, Baton Rouge: 1961.

———. "The Kuhnian Paradigm and the Darwinian Revolution in Natural History." In D. Roller, ed., *Perspectives in the History of Science and Technology*. Norman, OK: University of Oklahoma Press, 1971.

Hummel, Charles E. *The Galileo Connection*. Downer's Grove, IL: InterVarsity Press, 1986.

Hyers, Conrad. *The Meaning of Creation*. Atlanta: John Knox Press, 1984.

Lindberg, David C. and Ronald L. Numbers. *God and Nature: Historical Essays on the Encounter Between Christianity and Science*. Berkeley: University of California Press, 1986.
Mayr, Ernst. *The Growth of Biological Thought*. Cambridge: Harvard University Press, 1982.
Ratzsch, Del. *Philosophy of Science: The Natural Sciences in Christian Perspective*. Downer's Grove, IL: InterVarsity Press, 1986.
Van Till, Howard J. *The Fourth Day*. Grand Rapids, MI: Eerdmans, 1986.

### 5. Perspectives on Genesis 1

Blocher, Henri. *In the Beginning*. Downer's Grove, IL: InterVarsity Press, 1984
Burke, Derek, ed. *Creation and Evolution: When Christian's Disagree*. Leicester, UK: InterVarsity, 1985.
Fischer, Robert B. *God Did It, But How?* Grand Rapids, MI: Zondervan, 1981.
Houston, James M. *I Believe in the Creator*. Grand Rapids, MI: Eerdmans, 1980.
Hummel, Charles E. *The Galileo Connection*. Downer's Grove, IL: InverVarsity Press, 1986.
Hyers, Conrad. *The Meaning of Creation*. Atlanta: John Knox Press, 1984.
Pun, Pattle P. T. "A Theology of Progressive Creationism," *Perspectives on Science and Christian Faith* 39 (1987):9–19.
Van Till, Howard J. *The Fourth Day*. Grand Rapids, MI: Eerdmans, 1986.
Wiester, John. *The Genesis Connection*. Nashville: Thomas Nelson, 1983.
Young, Davis. *Christianity and the Age of the Earth*. Grand Rapids, MI: Zondervan, 1982.

### 6. The Origin of Life

Dubos, René. *Pasteur and Modern Science*. Garden City, NY: Doubleday, 1960.
Futuyma. Douglas J. *Science on Trial: The Case for Evolution*. New York: Pantheon Books, 1983.
Mader, Sylvia S. *Biology: Evolution, Diversity and the Environment*. Dubuque, IA: W. C. Brown, 1985.

Mayr, Ernst. *The Growth of Biological Thought*. Cambridge: Harvard University Press, 1982.
Norstog, K. and A. J. Meyerriecks. *Biology*, 2d ed. Columbus, OH: C. E. Merrill, 1983.
Pitman, Michael. *Adam and Evolution*. London: Rider Press, 1984.
Ruse, Michael. *Darwinism Defended*. Reading, MA: Addison-Wesley, 1982.
Thaxton, C. B., W. L. Bradley, and R. L. Olsen. *The Mystery of Life's Origin: Reassessing Current Theories*. New York: Philosophical Library, 1984.
Wiester, John. *The Genesis Connection*. Nashville: Thomas Nelson, 1983.

### 7. The Darwinian Revolution

Blocher, Henri. *In the Beginning*. Downer's Grove, IL: InterVarsity Press, 1984.
Bube, Richard H. *The Human Quest*. Waco, TX: Word Books, 1971.
Burke, Derek, ed. *Creation and Evolution: When Christians Disagree*. Leicester, UK: InverVarsity, 1985.
Denton, Michael. *Evolution: A Theory in Crisis*. Bethesda, MD: Adler and Adler, 1985.
Fischer, Robert B. *God Did It, But How?* Grand Rapids, MI: Zondervan, 1981.
Greene, John C. *Darwin and the Modern World View*. Baton Rouge: Louisiana State University Press, 1961.
Keeton, W. T. and J. L. Gould. *Biological Science*, 4th ed. New York: W. W. Norton, 1986.
Lindberg, David C. and Ronald L. Numbers. *God and Nature: Historical Essays on the Encounter Between Christianity and Science*. Berkeley: University of California Press, 1986.
Mayr, Ernst. *The Growth of Biological Thought*. Cambridge: Harvard University Press, 1982.
Mayr, Ernst. *Towards a New Philosophy of Biology*. Cambridge, MA: Harvard University Press, 1988.
Pitman, Michael. *Adam and Evolution*. London: Rider Press, 1984.
Rifkin, Jeremy. *Algeny*. New York: Viking Press, 1983.
Ruse, Michael. *Darwinism Defended*. Reading, MA: Addison-Wesley, 1982.
Thurman, L. Duane. *How to Think About Evolution*. Downer's Grove, IL: InverVarsity Press, 1978.
Van Til, Howard J. *The Fourth Day*. Grand Rapids, MI: Eerdmans, 1986.

Wiester, John. *The Genesis Connection.* Nashville, TN: Thomas Nelson, 1983.

Young, Davis. *Christianity and the Age of the Earth.* Grand Rapids, MI: Zondervan, 1982.

**8. Where Are You, Adam?**

Blocher, Henri. *In the Beginning.* Downer's Grove, IL: InterVarsity Press, 1984.

Futuyma, Douglas J. *Science on Trial: The Case for Evolution.* New York: Pantheon Books, 1983.

Gould, Stephen Jay. *Ever Since Darwin: Reflections in Natural History.* New York: American Museum of Natural History, 1977.

Greene, John C. *The Death of Adam.* Ames, IA: Iowa State University Press, 1959.

Houston, James M. *I Believe in the Creator.* Grand Rapids, MI: Eerdmans, 1980.

Hummel, Charles E. *The Galileo Connection.* Downer's Grove, IL: InterVarsity Press, 1986.

Mayr, Ernst. *The Growth of Biological Thought.* Cambridge, MA: Harvard University Press, 1982.

Pitman, Michael. *Adam and Evolution.* London: Rider Press, 1984.

Ruse, Michael. *Darwinism Defended.* Reading, MA: Addison-Wesley, 1982.

Wiester, John. *The Genesis Connection.* Nashville, TN: Thomas Nelson, 1983.

**9. Stewards of Creation**

Blocher, Henri. *In the Beginning.* Downer's Grove, IL: InterVarsity Press, 1984.

Elsdon, Ron. *Bent World.* Downer's Grove, IL: InterVarsity Press, 1981.

Granberg-Michaelson, Wesley. *A Worldly Spirituality.* San Francisco: Harper & Row, 1984.

Houston, James M. *I Believe in the Creator.* Grand Rapids, MI: Eerdmans, 1980.

Walsh, Brian, J. and J. Richard Middleton. *The Transforming Vision: Shaping a Christian World View.* Downer's Grove, IL: InterVarsity Press, 1984.

Wilkinson, Loren. *Earthkeeping: Christian Stewardship of Natural Resources.* Grand Rapids, MI: Eerdmans, 1980.

Wolters, Albert M. *Creation Regained: Biblical Basics for a Reformational Worldview*. Grand Rapids, MI: Eerdmans, 1985.

Wright, Richard T. "Responsibility for the Ecological Crisis," *Bioscience* 20 (1970): 851–853.

### 10. The Biomedical Revolution

Coppenger, Mark. *Bioethics: A Casebook*. Englewood Cliffs, NJ: Prentice-Hall, 1985.

Hill, A. V. *The Ethical Dilemma of Science*. New York: Rockefeller University Press, 1960.

Holmes, Arthur H. *Ethics: Approaching Moral Decisions*. Downer's Grove, IL: InterVarsity Press, 1984.

Jones, D. Gareth. *Brave New People*, rev. ed. Grand Rapids, MI: Eerdmans, 1985.

Jones, D. Gareth. *Manufacturing Humans*. Leicester, G.B.: Inter-Varsity Press, 1987.

Levine, Carol, ed. *Taking Sides: Clashing Views on Controversial Bio-Ethical Issues*. Guilford, CT: Dushkin Publishing Group, 1984.

Nelson, James B. and JoAnne S. Rohricht. *Human Medicine: Ethical Perspectives on Today's Medical Issues*. Minneapolis, MN: Augsburg Publishing House, 1984.

Shannon, Thomas A., ed. *Bioethics*. Mahwah, NJ: Paulist Press, 1980.

Varga, Andrew C. *The Main Issues in Bioethics*. Mahway, NJ: Paulist Press, 1980.

Wennberg, Robert N. *Life in the Balance: Exploring the Abortion Controversy*. Grand Rapids, MI: Eerdmans, 1985.

### 11. The Genetic Revolution

Cavalieri, Liebe F. *The Double-Edged Helix*. New York: Columbia University Press, 1981.

Cook, Harry. "The Biology Business," *Journal of the American Scientific Affiliation*, 34 (1982): 129–134.

Coppenger, Mark. *Bioethics: A Casebook*. Englewood Cliffs, NJ: Prentice-Hall, 1985.

Herrmann, Robert L. and J. M. Templeton. "The Vast Unseen and the Genetic Revolution," *J.A.S.A.* 37 (1985): 132–141.

Herrmann, Robert L., ed. *Making Whole Persons: Ethical Issues in Biology and Medicine*. J.A.S.A., 1980.

Jones, D. Gareth. *Brave New People*, rev. ed. Grand Rapids, MI: Eerdmans, 1985.
Levine, Carol. *Taking Sides: Clashing Views on Controversial Bioethical Issues*. Guilford, CT: Dushkin Publishing Group, 1984.
Mackay, Donald M. *Human Science and Human Dignity*. Downer's Grove, IL: InverVarsity Press, 1979.
Rifkin, Jeremy, *Algeny*. New York: Viking Press, 1983.
Teich, A. H., M. A. Levin, and J. H. Pace. *Biotechnology and the Environment*. Washington, DC: American Association for the Advancement of Science, 1985.
Wilcox, David L. "A Christian Integrative Framework for Biology," *Christian Scholar's Review* 12 (1983): 339–348.

### 12. The Environmental Revolution

Anonymous. *The Global Possible: Resources, Development and the New Century*. Washington, D.C.: World Resources Institute, 1984.
Brown, Lester et al. *State of the World* 1987. New York: W. W. Norton, 1987.
Chiras, Daniel D. *Environmental Science: A Framework for Decision Making*. Menlo Park, CA: Benjamin/Cummings, 1985.
Miller, G. Tyler. *Living in the Environment*, 4th ed. Belmont, CA: Wadsworth, 1985.
Rolston III, Holmes. *Philosophy Gone Wild: Essays in Environmental Ethics*. Buffalo, NY: Prometheus Books, 1986.
Sider, Ronald J. *Rich Christians in an Age of Hunger*, 2nd ed. Downer's Grove, IL: InterVarsity Press, 1984.
Southwick, Charles H. *Global Ecology*. Sunderland, MA: Sinauer Associates, 1985.
Wilkinson, Loren. *Earthkeeping: Christian Stewardship of Natural Resources*. Grand Rapids, MI: Eerdmans, 1980.

### 13. When Earth and Heaven Are One

Granberg-Michaelson, Wesley. *A Worldly Spirituality*. New York: Harper & Row, 1984.
Snyder, Howard. *A Kingdom Manifesto*. Downer's Grove, IL: InterVarsity Press, 1985.
Walsh, B. J. and J. R. Middleton. *The Transforming Vision*. Downer's Grove, IL: InterVarsity Press, 1984.

Wilkinson, Loren. *Earthkeeping: Christian Stewardship of Natural Resources*. Grand Rapids, MI: Eerdmans, 1980.
Wolters, Albert M. *Creation Regained*. Grand Rapids, MI: Eerdmans, 1985.
Wolterstorff, Nicholas. *Until Justice and Peace Embrace*. Grand Rapids, MI: Eerdmans, 1983.

# INDEX